GCSE

Philip Allan Updates, an imprint of Hodder Education, an Hachette UK company, Market Place, Deddington, Oxfordshire OX15 0SE

Orders

Bookpoint Ltd, 130 Milton Park, Abingdon, Oxfordshire OX14 4SB
tel: 01235 827720 fax: 01235 400454 e-mail: uk.orders@bookpoint.co.uk

Lines are open 9.00 a.m.–5.00 p.m., Monday to Saturday, with a 24-hour message answering service. You can also order through our website: www.philipallan.co.uk

ISBN 978-1-4441-0184-3

First published in 2005 as *Flashrevise Cards*

Impression number 5 4 3 2 1
Year 2014 2013 2012 2011 2010 2009

Printed in Spain

Hachette UK's policy is to use papers that are natural, renewable and recyclable products and made from wood grown in sustainable forests. The logging and manufacturing processes are expected to conform to the environmental regulations of the country of origin.

P01559

Physical geography

1 Types of rocks
2 Rock weaknesses
3 Weathering
4 Carboniferous limestone landscapes
5 Tectonic plate boundaries
6 Volcanoes
7 Impact of volcanoes on people
8 Earthquakes
9 Impact of earthquakes on people
10 Fold mountains
11 Water (hydrological) cycle
12 Human impacts on the water cycle
13 Drainage basins
14 River long profile
15 River landforms in the uplands
16 River landforms in the lowlands
17 Meanders and oxbow lakes
18 Deltas
19 River floods
20 River basin management
21 Example of a river
22 Processes and landforms of coastal erosion
23 Cliffs, caves, arches and stacks
24 Longshore drift
25 Landforms of coastal deposition
26 Spits
27 Methods of coastal protection
28 Coastal management
29 Example of crumbling cliffs
30 Processes and landforms of glacial erosion
31 Arêtes and corries (cirques)
32 Glaciated valleys
33 Landforms of glacial deposition
34 Weather and climate
35 Climate graphs
36 Anticyclones and depressions
37 Relief rainfall
38 Frontal depressions
39 Convectional rainfall
40 Tropical cyclones (hurricanes)
41 Natural hazards
42 Ecosystems
43 Tropical rainforests
44 Rainforest deforestation
45 Sustainable development
46 Desertification
47 Pollution
48 Acid rain
49 The greenhouse effect
50 Global warming

Human geography

Types of rocks

Q1 What is igneous rock? Name an example.

Q2 What is sedimentary rock? Name an example.

Q3 How are metamorphic rocks different from igneous and sedimentary rocks? Name an example of a metamorphic rock.

Q4 What is meant by the 'economic use' of a rock?

ANSWERS

all rocks are one of three types — igneous, sedimentary or metamorphic

A1 rock formed by volcanic activity; examples are granite and basalt

A2 rock formed from sediments such as sand and mud laid down on the sea bed; examples are chalk, limestone, sandstone and clay

A3 metamorphic rock is igneous or sedimentary rock altered by heat and pressure; examples are marble (from limestone) and slate (from clay)

A4 the ways in which a rock can be profitably used or sold, e.g. limestone is used as building stone, fertiliser (lime) and for making cement

examiner's note The type of rock is important because it affects rates of erosion, water storage and soil fertility.

Rock weaknesses

Q1 Name rock weaknesses A, B and C.

Q2 What has happened to rock layer X?

Q3 State two differences between weaknesses B and C.

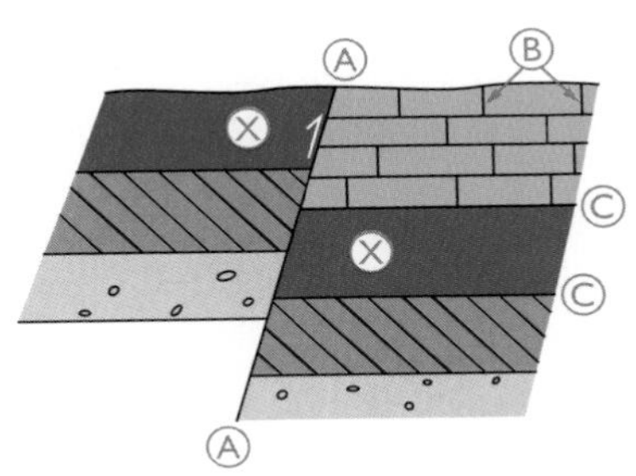

ANSWERS

weak points within and between layers of rock

A1 A = fault; B = joint; C = bedding plane

A2 it was once a continuous layer (or bed) of rock; faulting has broken it into two layers at different heights; the layer on the left has been moved up by earth movements

A3
- joints (B) are weaknesses within the same layer of rock; bedding planes (C) are weaknesses between different layers (or beds) of rock
- joints are vertical weaknesses; bedding planes are horizontal weaknesses

***examiner's* note** Rock weaknesses are important because they affect rates of weathering and erosion. The greater the number of weaknesses, the greater the speed at which rocks are broken down and carried away.

Weathering

Q1 Name the three types of weathering.

Q2 What type of weathering is freeze–thaw?

Q3 Explain how freeze–thaw weathering works.

Q4 How is erosion different from weathering?

A1 physical (or mechanical), chemical and biological

A2 physical

A3 when temperatures fall below 0°C, water trapped in cracks in the rock freezes; as water freezes, it expands and puts pressure on the surrounding rock; when temperatures rise above 0°C, thawing releases the pressure; after freezing and thawing happens many times, pieces of rock break off

A4 movement is needed for erosion to occur; erosion is the wearing away of the land by rivers, waves and glaciers

***examiner's* note** Weathering happens fastest when rocks are full of weaknesses (joints, bedding planes and faults) and extreme weather changes occur (e.g. great heat and extreme cold).

Carboniferous limestone landscapes

Q1 Name landforms A–F.

Q2 Describe what happens to the river between points 1 and 2.

Q3 Why does this happen?

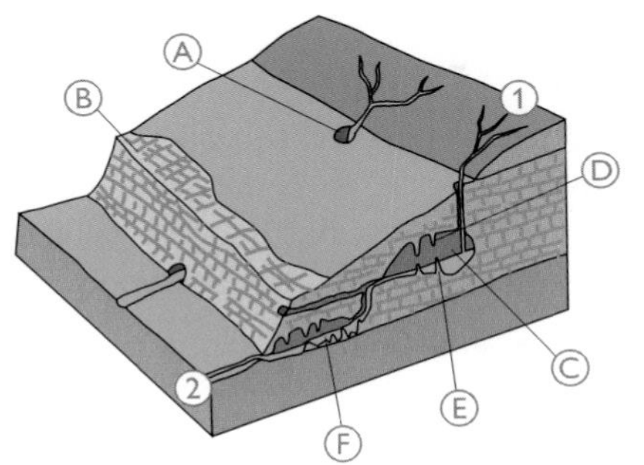

ANSWERS

A1 A = swallow hole; B = limestone pavement; C = cavern; D = stalactite; E = stalagmite; F = cave

A2 at first the river flows on the surface; then it disappears underground down a swallow hole; next it flows underground through caves; finally it reappears and flows on the surface

A3 the river flows on the surface in places where rocks are impervious; it flows underground through Carboniferous limestone because this rock is full of holes; the holes are widened by limestone solution

examiner's note Karst landforms are created by limestone solution, a type of chemical weathering. Carboniferous limestone rock dissolves in rainwater which has been made acidic by the presence of carbon dioxide in the atmosphere.

Tectonic plate boundaries

Q1 What is a plate boundary?

Q2 Name the three types of plate boundary (A, B and C) in the diagram.

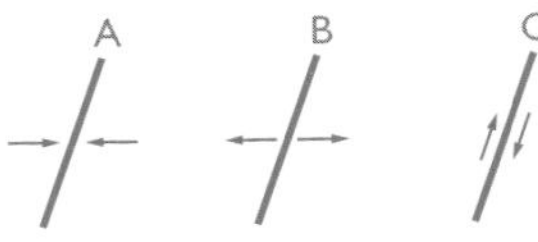

Q3 Why do earthquakes occur along all three types of plate boundary?

Q4 Why do volcanoes not form along plate boundary C?

ANSWERS

meeting point of two of the enormous pieces of rock that make up the Earth's crust

A1 a place where two great pieces of crustal rock meet

A2 A = destructive or convergent; B = constructive or divergent; C = conservative or passive

A3 movement occurs from time to time along all plate boundaries; earth movements create shock waves and the earth shakes (or quakes)

A4 a volcano forms where there is an outlet for magma from the interior of the Earth to the surface; along plate boundary C the two plates are simply moving past each other and no such outlets are formed

***examiner's* note** Plate boundaries are the most unstable areas of the Earth's crust. The risk of earthquakes and volcanic activity is highest here.

Volcanoes

Q1 Describe the differences between volcanoes A and B in terms of height, shape and composition.

Q2 Explain why volcanoes A and B form along different types of plate boundary.

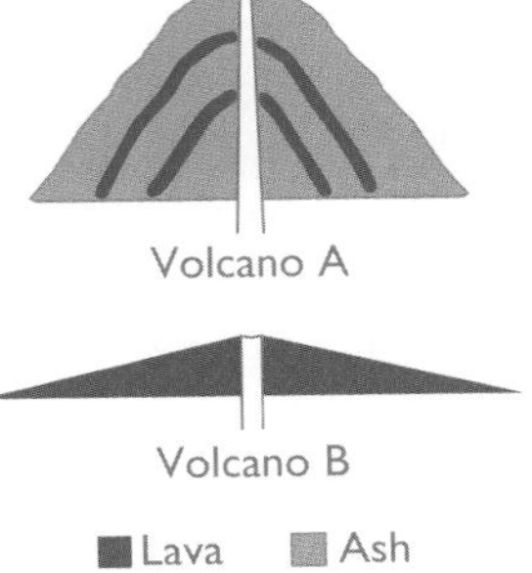

ANSWERS

A1
- height: volcano A is taller than volcano B
- shape: volcano A is a steep-sided cone whereas volcano B has gentler slopes and a wider base
- composition: volcano A is a composite cone of lava and ash whereas volcano B is lava only

A2 volcano A forms at destructive plate boundaries — as one plate is destroyed, the crust melts and magma rises; volcano B forms at constructive plate boundaries — molten magma from the Earth's interior fills the gap between plates as they move apart

***examiner's* note** When a question asks for differences, use comparatives such as 'taller than', 'wider than' or link the two parts of your answer with 'whereas'.

Impact of volcanoes on people

Q1 Identify three short-term negative impacts of volcanoes on people.

Q2 Identify four long-term positive impacts of volcanoes on people.

Q3 Why do volcanoes kill fewer people than earthquakes?

Q4 Give reasons why more than half a billion people live in areas with active volcanoes.

the effects, both good and bad, on people living in areas of volcanic activity

A1 loss of life; buildings and farmland destroyed; public services disrupted

A2 useful minerals, e.g. sulphur; tourist attraction; fertile volcanic soils; geothermal power

A3 earthquakes happen without warning, but warning signs from volcanoes include rising temperatures in the crater and minor eruptions; volcanoes can therefore be monitored and people have time to move out of the way of lava flows when an eruption is predicted

A4 lava and ash weather to make extremely fertile soils — some of the best in the world for farming; see also A2 above

examiner's note When the question asks for impacts, remember to look for both good and bad/positive and negative.

Earthquakes

Q1 What is the difference between the focus and the epicentre of an earthquake?

Q2 Explain why strong earthquakes form along destructive plate boundaries.

Q3 Where will the earthquake be strongest: at E1, E2 or E3? Why?

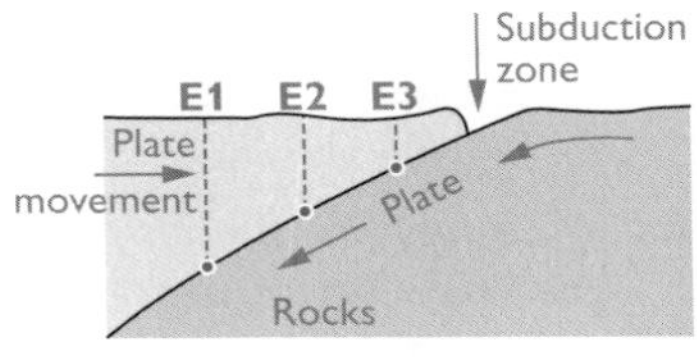

ANSWERS

vibrations that shake the ground, mostly in areas close to plate boundaries

A1 the focus is the underground source of the earthquake; the epicentre is the point on the surface above the focus where shaking is greatest; shock waves radiate out from the epicentre in a circular pattern, becoming less strong with increasing distance from the centre

A2 one rock plate is sinking (subducting) and being destroyed; forced movement of rock against rock creates great friction; friction causes the vibrations that reach the surface above

A3 at E3, because the focus is closest to the surface; the distance for the shock waves to travel is less, so less of their strength is absorbed and lost

examiner's note Earthquakes are very common; there are more than 6000 per year worldwide, although only about 20 are strong enough to cause major damage.

Impact of earthquakes on people

A tale of two earthquakes

Date: 22 December 2003
Place: California, USA
Richter scale: 6.5
Damage: clock tower toppled, killing three people

Date: 26 December 2003
Place: Bam, Iran
Richter scale: 6.5
Damage: 60% of buildings flattened, thousands killed and many more injured

Q1 Look at the descriptions of the two earthquakes. What are the similarities?

Q2 What are the differences?

Q3 Explain the differences.

A1 the strength of the earthquakes; the date they occurred

A2 the numbers killed and amount of damage caused: these were so much greater in Bam

A3 differences in design of buildings and preparedness for earthquakes: in California, strict building regulations are enforced, tall buildings have steel supports and deep reinforced foundations, and regular earthquake practice drills are held; in Iran, houses were built of mud bricks, building regulations were ignored by builders and corrupt officials, and it is a much poorer country with fewer resources for earthquake protection

***examiner's* note** Immediate impacts of earthquakes are deaths, injuries and destruction of buildings, roads and bridges. Long-term impacts are homelessness, poverty and high economic costs of recovery.

Fold mountains

Q1 Name two young fold mountain ranges.

Q2 Why do young fold mountains form the world's highest peaks?

Q3 Why are they formed at destructive (convergent) plate boundaries?

Q4 Name the economic activities found in high mountain areas.

ANSWERS

high mountain ranges formed by the upfolding of sediments during powerful earth movements

A1 examples include the Rockies, Andes, Himalayas and Alps

A2 they were formed recently and some are still being formed; there has not been enough time for them to have been eroded away

A3 two massive rock plates collide here; sediments on the sea bed between the plates are compressed and forced upwards in a series of great folds; many fold mountains contain volcanoes, which increase their height

A4 hydroelectric power (HEP) and water storage; animal farming, forestry and mining; tourism, especially skiing

examiner's note Earthquakes, volcanoes and fold mountain ranges all follow the line of destructive plate boundaries.

Water (hydrological) cycle

Q1 Name processes A–H.

Q2 Why is this series of processes called a cycle?

Q3 What is an aquifer?

Q4 How can people obtain water supplies from aquifers?

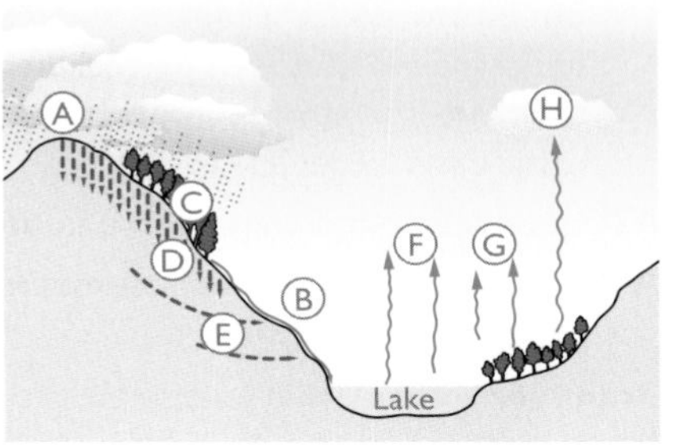

ANSWERS

movement of water from the atmosphere onto the Earth's surface and back into the atmosphere

A1 A = precipitation; B = runoff; C = interception; D = infiltration; E = groundwater flow; F = evaporation; G = evapotranspiration; H = condensation

A2 it occurs continuously: water is constantly being recycled; water can be stored for a time in surface stores such as lakes or underground stores in permeable rocks; eventually it is evaporated back into the atmosphere where condensation and precipitation restart the cycle

A3 an underground water store in permeable rock (e.g. chalk)

A4 normally by sinking wells; sometimes water reaches the surface naturally in springs

examiner's note Fresh water is the vital natural resource for life on Earth; natural water stores are vital for supplying cities and growing crops.

ANSWERS

Human impacts on the water cycle

Q1 How does the amount of runoff change after deforestation?

Q2 Explain why it changes.

Q3 Why is runoff faster from urban areas than from farmland?

Q4 Are the actions of people leading to more frequent and more severe river floods? Explain your answer.

ANSWERS ››

changes to water-cycle processes caused by the activities of people

A1 runoff increases

A2 without trees none of the precipitation is intercepted; runoff is not blocked by tree stems, especially on steep slopes; water is no longer lost to the atmosphere by transpiration

A3 most farmland has surface vegetation which intercepts rain and soils which store water; hard surfaces like roads in urban areas stop water from infiltrating underground; rain is channelled down gutters, into drains and straight back into rivers

A4 by increasing and speeding up runoff, humans are making natural floods worse

examiner's note Runoff and floods have natural causes, but they are increasing because of human activities.

Drainage basins

Q1 Name drainage-basin features A–F.

Q2 What is the difference between a tributary and a distributary?

Q3 Explain where the highest land is found.

Q4 Explain where the lowest land is found.

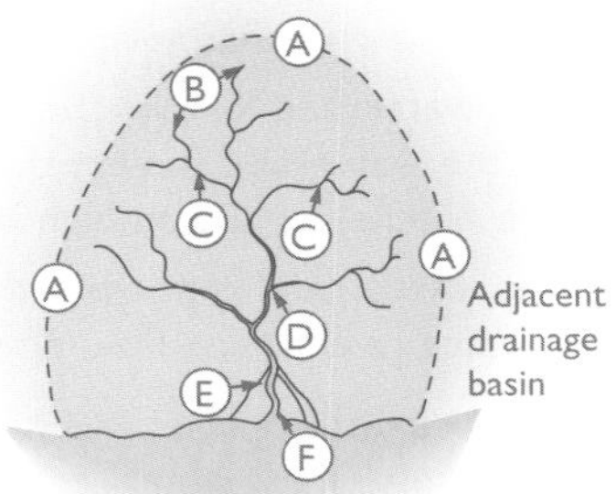

the area of land drained by a river and its tributaries

A1 A = watershed; B = source; C = tributary; D = confluence; E = distributary; F = mouth

A2 a tributary is a small stream or river that flows into a larger one; a distributary is a small river channel carrying water into the sea after the river has split up in a delta

A3 the watershed follows the highest land; rivers on the other side of the watershed flow in the opposite direction; the sources of the tributaries are on high land just below the watershed

A4 the lowest land is at the mouth where the river meets the sea; surface rivers must flow downhill

examiner's note All the precipitation that falls on a drainage basin eventually reaches the main river and runs into the sea.

ANSWERS

River long profile

Q1 State two differences between the profile at point A and at point B.

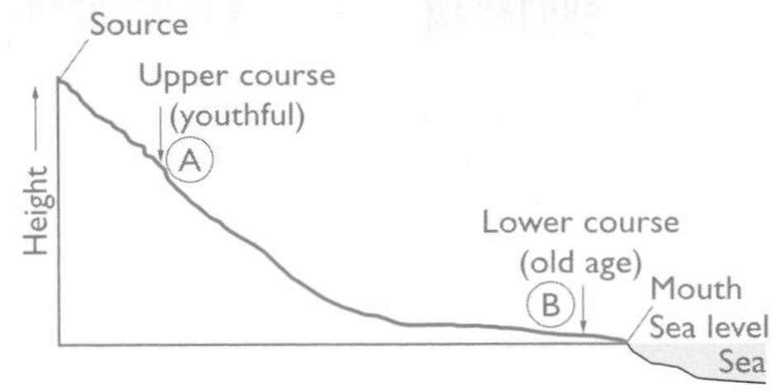

Q2 Why does the shape of the long profile change between points A and B?

Q3 Name the main landforms found in the upper and lower courses.

ANSWERS

summary line for the shape and gradient of a river bed from source to mouth

A1 • steep at A; gentle at B
• irregular at A; smooth at B

A2 at A the river is cutting down by vertical erosion; at B the river is cutting sideways by lateral erosion, because it has almost reached sea level

A3 • upper course: steep-sided, V-shaped valley; interlocking spurs; gorge; waterfall; rapids
• lower course: meander; oxbow lake; levée; floodplain; delta

***examiner's* note** The long profile is the view up and down the valley; the cross profile is the view across the valley, which for a river valley is a V-shape.

River landforms in the uplands

Q1 Describe the river and channel features shown in the photo.

Q2 Explain how the channel features in the photo were formed.

ANSWERS

formed mainly by vertical erosion as the river cuts down into the land

A1 small waterfalls and rapids; rocky bed with bands of hard rock; step-like channel; fast-flowing river; brown water (peaty)

A2 hard rock outcrops are difficult to erode; areas of softer rock between them are easier to erode; water falls over the hard rock outcrops to the lower, eroded areas; rivers cut down by vertical erosion; abrasion and corrosion are examples of the processes of erosion

examiner's note In photo questions, pay great attention to the command word. 'Describe' means write about what can be seen on the photo. 'Explain' in this case means use knowledge about landform formation.

River landforms in the lowlands

Q1 Name landforms A–E.

Q2 Explain the formation of the floodplain.

Q3 Which one of the landforms A–E is formed only by deposition?

Q4 How is landform E different in big rivers such as the Nile, Mississippi and Ganges?

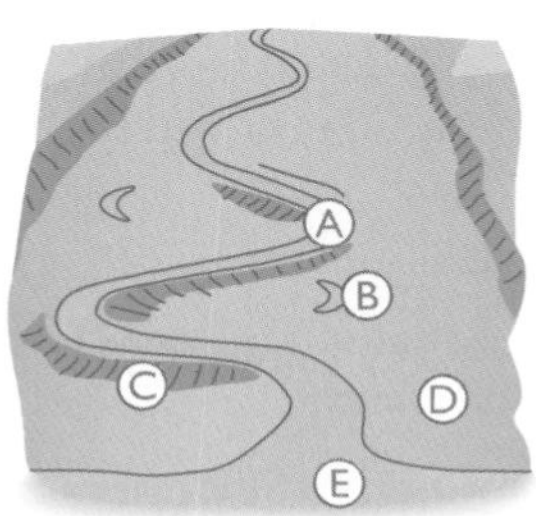

A1 A = meander; B = oxbow lake; C = levée; D = floodplain; E = estuary

A2 the floodplain is formed by both deposition and lateral erosion; every time the river floods it deposits a layer of silt, which accumulates over time into a great thickness of silt; the floodplain is widened by lateral erosion on the outside bends of meanders; the sides of the floodplain are cut back where outside bends reach them

A3 C — levée

A4 deltas form; instead of one channel, water enters the sea by many channels

***examiner's* note** Lateral (sideways) erosion is dominant in the lowlands because the river is near to sea level and cannot cut down any more. Its large load of silt is deposited at points where the water is slow flowing.

Meanders and oxbow lakes

Q1 What is the difference between the river cliff and the slip-off slope?

Q2 How and why do oxbow lakes form?

Q3 Where is the deepest water in the channel of a meandering river?

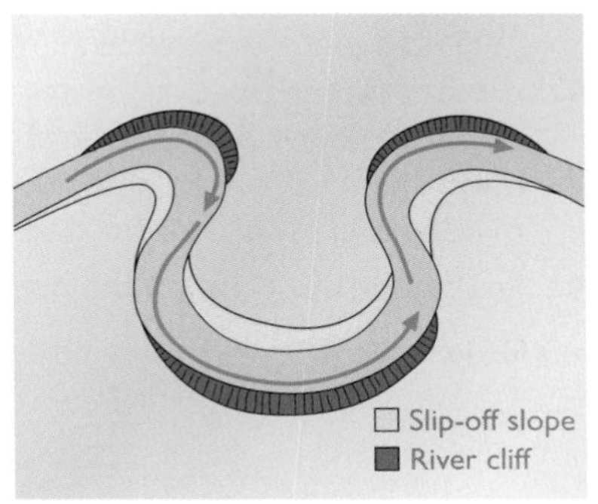

common landforms where rivers flow across flat land in the lowlands

A1 the river cliff is the steep bank on the outside bend of a meander, formed by erosion; the slip-off slope is a gentle slope of sand or stones on the inside bend, formed by deposition

A2 erosion causes the outside bends of two meanders to come closer together; during flood the river may cut across the narrow neck and flow straight; the old meander loop is cut off to form the oxbow lake; silt is deposited to seal the lake off from the river

A3 just below the bank on the outside bend

***examiner's* note** Remember the basics: river cliff = erosion, outside bend; slip-off slope = deposition, inside bend.

Deltas

Q1 Explain the formation of a delta.

Q2 Why do most big rivers have deltas?

Q3 Why are deltas challenging places for people to live?

Q4 Some of the world's most densely populated rural areas are in deltas (e.g. the Ganges delta in Bangladesh). Explain why.

ANSWERS

triangular-shaped areas of flat land where a river splits into many different channels (distributaries)

A1 rivers have large loads and are slow flowing near the sea; meeting denser salt water holds back their flow; they split into separate channels, depositing silt over a wide area

A2 big rivers have large loads of sediment, e.g. the Mississippi, the Ganges

A3 large areas of swamp and standing water make transport difficult; land needs to be drained and reclaimed for farming; high flood risk from heavy rains and rising sea levels

A4 the soils are incredibly fertile; fertility is renewed by regular flooding

examiner's note People living in deltas will be the first to be affected by rising sea levels from global warming.

River floods

Q1 State three natural causes of river floods.

Q2 Name two river landforms formed during floods.

Q3 Describe the possible impacts of flooding on people.

Q4 Why are more people being affected by flooding in the UK than in the past?

when rivers break their banks, covering surrounding land areas that are usually dry

A1 wet season with prolonged and heavy rainfall; summer melting of snow and ice in the mountains; torrential downpours such as thunderstorms or rainfall during cyclones

A2 floodplain, levée and oxbow lake

A3 people drowned; houses flooded and made uninhabitable; bridges and roads washed away; crops ruined and livestock drowned; soils made from silt are very fertile

A4 more houses have been built in high-risk locations on floodplains; more urban areas which increase runoff

examiner's note River flooding is normal. The immediate effects are negative, but long-term effects can be good for farmers.

River basin management

Q1 Name three methods used to reduce river flooding.

Q2 Describe how they work.

Q3 How sustainable are these methods?

Q4 What can be done in rural areas to reduce runoff into streams and rivers?

the attempt by people to control rivers for their own purposes

A1 dams, embankments and walls in urban areas

A2
- dams hold back water for controlled release
- embankments increase the height of the river banks
- walls in urban areas hold back water and stop the river from changing its course

A3 sustainability is poor; high cost of maintenance; interference with the river's natural flow; silt trapped in dams and channel beds

A4 plant trees which increase rates of interception, hold back the flow of rainwater and increase the rate of infiltration — this is more sustainable

***examiner's* note** Rivers have many uses which means large numbers of people live close to them. For this reason many rivers are managed.

Example of a river

Q1 Describe the physical features shown in the photo.

Q2 What could happen next to the course of the river?

Q3 When in flood, how much of the land in the photo will be under water?

ANSWERS

A1 large, wide meander loops; flat land (floodplain) on both sides of the river; land rising up gently from the valley floor

A2 the meander loop could become narrower and be cut through by the river; an oxbow lake would be left in the place of the present meander loop

A3 flat land next to the river will be flooded first; flooding can be expected on the floodplain; the land rises quite steeply on the valley sides and flooding is unlikely here

***examiner's* note** The River Cuckmere is now managed. Water is taken out of it further upstream and flooding is less likely than previously.

Processes and landforms of coastal erosion

Q1 What is abrasion (corrasion)?

Q2 When is wave erosion greatest?

Q3 Name the three other processes of coastal erosion.

Q4 Which of the following landforms are formed by coastal erosion?

- arch
- bar
- bay
- beach
- cave
- cliff
- headland
- spit
- stack
- wave-cut platform

the ways in which waves wear away the coastline and the resulting coastal features

A1 the wearing away of rocks by waves flinging pebbles against cliff faces

A2 during storms when waves are big and powerful, known as destructive waves; when winds blow onshore, driving the waves against the cliffs

A3 attrition, corrosion and hydraulic action

A4 arch, bay, cave, cliff, headland, stack and wave-cut platform (all the others are landforms of deposition)

***examiner's* note** In examination answers about landforms of erosion, always refer to the processes of erosion responsible for the formation of the landforms. Use a memory aid such as HAAC (hydraulic action, abrasion, attrition and corrosion) to help you remember the processes.

Cliffs, caves, arches and stacks

Q1 Name landforms A–D.

Q2 Which rock features aid erosion?

Q3 Explain how a stack is formed.

Q4 What is the name for the last piece of rock left after a stack has collapsed?

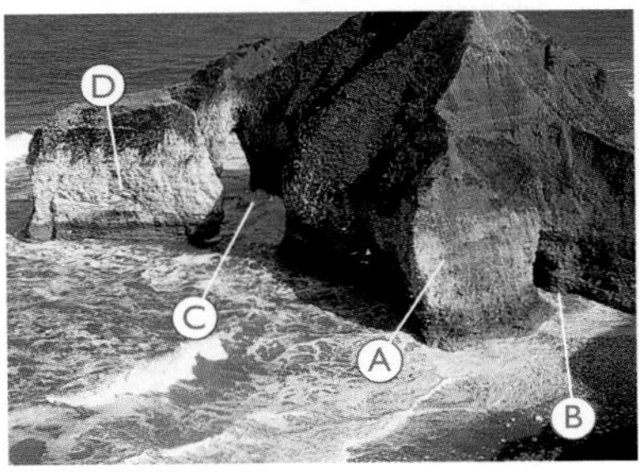

Flamborough Head

ANSWERS

landforms created by a sequence of erosion, ending with an isolated piece of rock (stack)

A1 A = cliff; B = cave; C = arch; D = stack

A2 rock weaknesses, both horizontal (bedding planes) and vertical (joints)

A3 waves attack a vertical weakness in the cliff; further erosion widens the crack into a cave; an arch is formed when the cave is eroded through the headland; the arch roof collapses after its base suffers from more wave erosion; an isolated pillar of rock (stack) is left

A4 stump

examiner's note Caves, arches and stacks are common in rocks full of weaknesses, e.g. chalk in the photo of Flamborough Head.

Longshore drift

Q1 What is the difference between swash and backwash?

Q2 Explain the direction of longshore drift shown in the diagram.

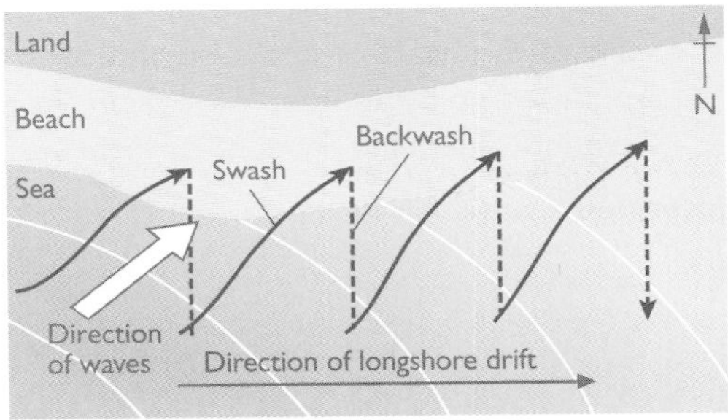

ANSWERS

the movement by waves of sand and pebbles along a coastline

A1 swash is the wave direction of breaking waves up the beach; backwash is the direction of water returning down the beach

A2 waves approach the beach from the southwest; once wave power is lost, pebbles roll back into the sea with the backwash at a right angle to the beach, in the same direction as the slope; pebbles are picked up by another wave from the southwest; repeated wave movements carry pebbles from west to east in this example

examiner's note Longshore drift is an important process; waves transport materials eroded from cliffs to sheltered locations where they are deposited to form beaches. Around Britain, longshore drift is from west to east along the coast of the English Channel, from south to north along the west coast of England, Wales and Scotland and from north to south down the east coast of Scotland and England.

Landforms of coastal deposition

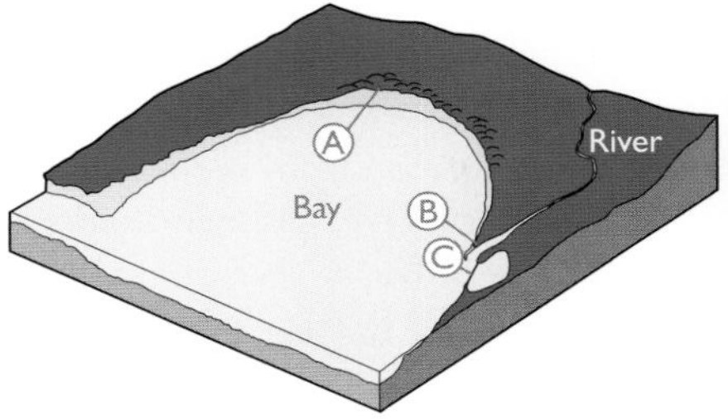

Q1 Name landforms A–C.

Q2 How are they different?

Q3 What is similar about their formation?

Q4 What is the evidence that shows the direction of longshore drift?

ANSWERS

made from sand, shingle and pebbles, carried and deposited by longshore drift

A1 A = beach; B = spit; C = bar

A2 in shape and location: beaches tend to be semi-circular, often in a bay; spits are narrow strips of land ending in the open sea, sometimes with a hooked end; bars are ridges which extend in a line across the mouth of a bay

A3 all are formed from sediments carried by longshore drift; deposition takes place in sheltered locations, such as bays or river mouths

A4 direction of the spit: longshore drift must be from left to right

***examiner's* note** Coastal deposition landforms form best in shallow waters where wave action is weak. Here, waves tend to be constructive, not destructive (as for erosion).

Spits

Q1 Describe the shape of the spit.

Q2 Explain its formation.

Q3 Why are extensive mud flats and salt marshes present?

Q4 Why do some spits stop growing or break?

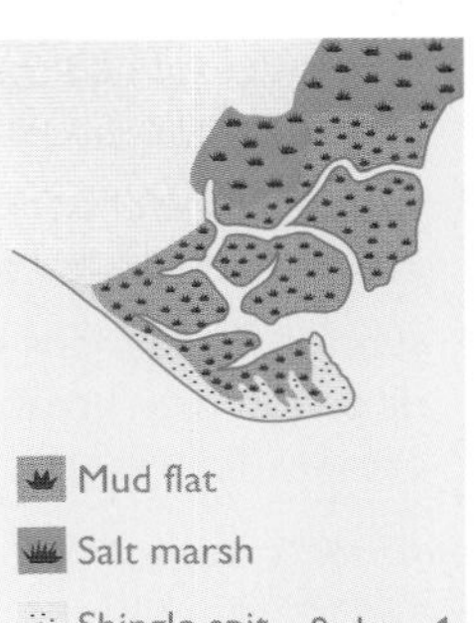

ANSWERS

ridges of sand and shingle attached to the land at one end and terminating in the sea

A1 narrow and straight for 1–2 km from the land, with four hooks at the sea end, the last one being the largest

A2 deposition begins at a bend in the coastline; longshore drift builds up a ridge of sediment out to sea; in the open sea currents rework the sediments into hooks

A3 the spit shelters the water behind it from currents; sediments accumulate, which plants colonise

A4 supply of sediment may be reduced; storm waves from a different direction may break the spit at a narrow point; strong currents prevent further growth out to sea

***examiner's* note** Note the different answers needed for questions 1 and 2, as the command word changes from 'describe' to 'explain'.

Methods of coastal protection

Q1 Name three methods of coastal protection.

Q2 Explain how they protect the coastline.

Q3 Where are they used most?

Q4 How sustainable are these types of coastal protection?

used to try to stop coastal erosion

A1 sea walls, blocks of rock and groynes

A2
- sea walls and blocks of rock work in the same way, breaking the force of the waves before they reach the cliffs
- groynes trap sediment, widen the beach, keep waves off the cliffs

A3 in seaside resorts and where homes need protection; where cliffs made of soft rock are crumbling, particularly along the south and east coasts of England

A4 low sustainability because of:
- high costs of construction and maintenance
- other unprotected parts of the coast suffer from increased erosion

examiner's note At a cost of over £10 million per kilometre, full protection is possible only where the economic need is greatest.

Coastal management

Q1 What are groynes and why are they a common sight in British seaside resorts?

Q2 Consider the following policies for dealing with rising sea levels.

- policy A: extend sea walls and build barrages
- policy B: managed retreat — do nothing and let the sea flood low-lying coastal areas

Decide who is in favour of each policy and say why.

A1 they are wooden (or metal) barriers running down the beach; they trap sand and shingle carried by longshore drift, increasing beach sizes

A2
- in favour of policy A: local authorities in urban areas; people who have homes or caravans on the coast; farmers — all will suffer economic losses as sea levels rise
- in favour of policy B: local authorities in rural areas — as they do not have the money for sea defences; environmentalists — to increase wildlife wetland habitats

examiner's note The threat of rising sea levels is causing a big debate about policies for coastal management. The number of people in favour of 'doing nothing' is increasing because of the cost and difficulty involved in battling against the power of the sea.

Example of crumbling cliffs

Q1 Describe the features in the photo that show the cliffs are crumbling.

Q2 What would be the chances of protecting these cliffs?

Q3 Why is the problem of crumbling cliffs in the UK becoming more serious?

ANSWERS

Walton-on-the-Naze in Essex, where the cliffs are built of soft sands

A1 rock, soil and vegetation are slipping down the cliffs; great mass of cliff debris looking unstable; cliff top is a long way inland from the beach

A2 it is unlikely that these cliffs can be protected for several reasons: impossibly expensive; cliffs are high; rain will loosen rocks and soil so that they slip more; difficult to hold back so much cliff material; narrow beach at bottom without a good site for building a sea wall

A3 the number of major storms seems to be increasing; sea levels are slowly rising due to global warming; places that were once inland are now in danger from future collapses

examiner's note Cliffs like these are a common sight on the east coast of England because soft rocks outcrop in many places.

Processes and landforms of glacial erosion

Q1 Name the two processes of glacial erosion.

Q2 Explain how these two processes operate.

Q3 Which of the following landforms are created by glacial erosion?

- arête
- drumlin
- horn (pyramidal) peak
- ribbon lake
- truncated spur
- corrie (cirque)
- hanging valley
- lateral moraine
- terminal moraine
- U-shaped valley

the ways in which glaciers wear away the land surface and the features that result

A1 abrasion and plucking

A2
- abrasion: rocks carried in the bottom of the ice work like giant files to erode ground surfaces
- plucking: ice freezes to rock surfaces; as the glacier moves, pieces of rock are pulled out and carried away

A3 arête, corrie (cirque), hanging valley, horn (pyramidal) peak, ribbon lake, truncated spur and U-shaped valley (the other three are landforms of glacial deposition)

***examiner's* note** The great weight and thickness of glaciers make them powerful agents of erosion. They sharpen peaks, deepen valleys and form many lakes in mountainous regions.

Arêtes and corries (cirques)

Q1 Describe the corrie shown in the diagram.

Q2 Explain how a corrie is formed.

Q3 Why are two or more corries needed for arêtes and horn peaks to form?

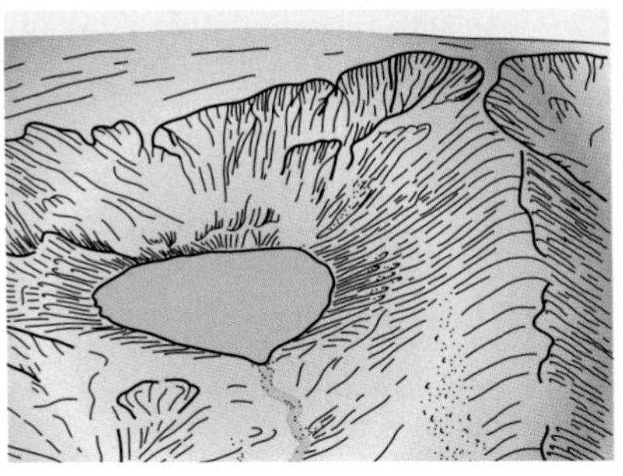

ANSWERS

the knife-edged peaks and steep circular hollows at the top of mountains

A1 steep-sided rocky hollow, circular in shape; round lake on its floor (tarn lake); steep slope on right side leading to a narrow ridge (arête); open front to corrie hollow with a rock lip

A2 frost action breaks off pieces of rock from the headwall; as the glacier moves, rock is plucked from the headwall; these loose rocks are used as tools for abrasion; erosion is strongest at the base of the headwall where the deepest hollow is made

A3 both landforms are formed by the headwalls of corries retreating by erosion and weathering, until only narrow pieces of rock are left between them

examiner's note Arêtes and corries are dramatic landforms seen at the tops of high mountain ranges.

Glaciated valleys

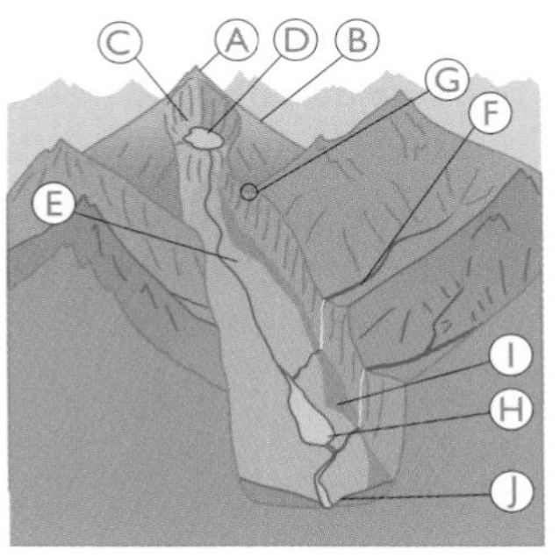

Q1 Name landforms A–J.

Q2 Describe and explain the shape of the cross profile of a glaciated valley.

Q3 Why is the long profile of a glaciated valley irregular?

ANSWERS

A1 A = horn peak; B = arête; C = corrie; D = tarn lake; E = U-shaped valley; F = hanging valley; G = truncated spur; H = ribbon lake; I = lateral moraine; J = terminal moraine

A2 U-shape because the glacier fills the whole valley; it erodes both the sides and the floor by abrasion as it flows downvalley

A3 the glacier exploits differences in rock hardness; soft rocks are abraded more quickly; hollows on the valley floor fill up, forming ribbon lakes after glaciers melt

examiner's note Glaciated valleys are the main areas for settlement, farming and transport routes in high mountain regions.

Landforms of glacial deposition

Q1 What is boulder clay?

Q2 Name the four types of moraine. What is the main difference between them?

Q3 How are drumlins different from moraines?

Q4 When and where do glaciers deposit?

ANSWERS

material left by glaciers, mainly boulder clay, found on valley floors and lowlands

A1 an unsorted mixture of clay and boulders of different sizes

A2
- ground, lateral, medial and terminal
- the main difference is location, e.g. lateral moraine occurs along the sides of the glacier whereas terminal moraine forms at the end of a glacier

A3 in drumlins boulder clay is moulded into egg-shaped mounds, usually found in clusters

A4 in the lowlands, where the glacier's carrying capacity is reduced by melting; at the snout of the glacier when it retreats

***examiner's* note** Boulder clay covers large areas in the Midlands and East Anglia and can be a good farming soil. At the coast, soft cliffs suffer rapid erosion.

Weather and climate

Q1 What is weather?

Q2 Why do farmers and supermarkets take note of weather forecasts?

Q3 How is climate different from weather?

Q4 Summarise the climate of the UK.

ANSWERS

factors including temperature, precipitation, sunshine and cloud, wind direction and speed

A1 the condition of the atmosphere at any given time, leading to day-to-day variations in such factors as temperature and rainfall

A2
- farmers: farm work such as planting, harvesting or bringing livestock indoors has to be planned according to the weather
- supermarkets: to predict what will sell, e.g. ice cream and meat for barbeques when hot weather is forecast

A3 climate is the average weather conditions recorded at a place over a long period of time, usually at least 30 years

A4 rain all year, warm summers (not hot) and mild winters (rarely very cold) — overall cool and wet

examiner's note Weather and climate have profound effects on people, affecting things such as food supply, leisure and recreation, water supply and transport.

Climate graphs

Q1 This graph shows Birmingham's climate. Describe the main features of temperature and precipitation.

Q2 Give a summary of Birmingham's climate.

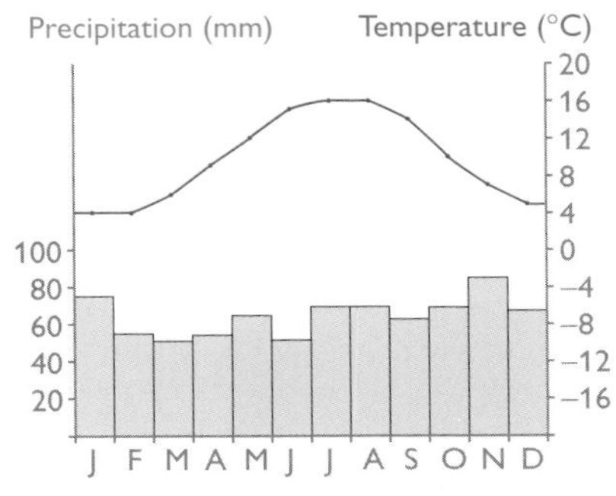

A1
- highest temperature 16°C in July and August; lowest 4°C in January and February; annual range of temperature 12°C
- most precipitation in November (84 mm); precipitation evenly distributed throughout the year, with no month receiving less than 50 mm

A2 warm, wet summers (15–16°C), cool, wet winters (4–5°C)

***examiner's* note** When interpreting any climate graph, always state highest temperature and month, lowest temperature and month, annual range of temperature, month with most precipitation and distribution during the year (season with most rain or rain all year). Average monthly temperature and total annual precipitation can also be calculated. Always state units of measurement.

Anticyclones and depressions

Q1 Why do anticyclones give mainly dry weather?

Q2 What are the differences in anticyclonic weather between summer and winter in the UK?

Q3 What types of weather are associated with depressions? Explain why.

Q4 Are there any differences in depression weather between summer and winter in the UK?

areas of high and low atmospheric pressure that control the weather

A1 air sinks in the centre of the high pressure; air currents are unable to rise high into the atmosphere; air is being warmed up by sinking instead of being cooled by rising

A2 hot, dry, sunny weather in summer can lead to heat waves; clear and cold weather in winter gives night frosts and fog

A3 cloudy, wet and windy weather; air rises in the centre of the low pressure; as the air rises, it cools, and moisture condenses to form cloud and rain

A4 weather is less different between seasons than for anticyclones; winds are often stronger during winter storms

examiner's note In the UK, depressions are more common than anticyclones and dominate the weather for most of the year.

Relief rainfall

Q1 Name the three types of rainfall.

Q2 This diagram shows the formation of relief rainfall and rain shadow. Write labels for what is happening at A–F on the diagram.

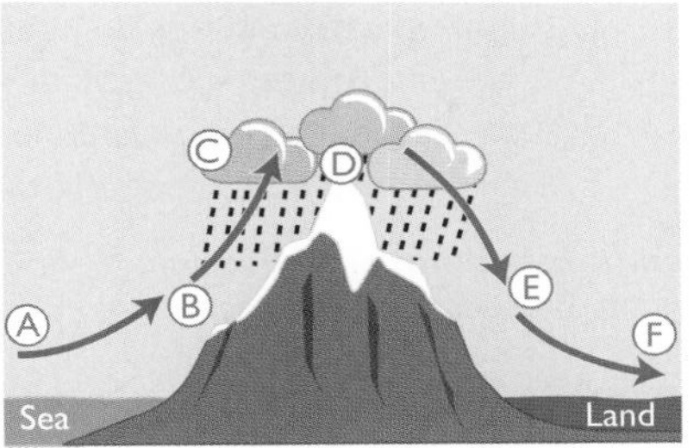

ANSWERS

precipitation caused by the cooling of moist air as it is forced to rise over mountains

A1 relief, frontal and convectional

A2 A = moist winds blowing onshore
B = air rises and cools
C = condensation forms a sheet of clouds
D = water droplets increase in size and rain falls
E = air descends and warms up
F = air becomes dry, forming the rain shadow

examiner's note All three types of rain are formed in the same way: by air rising and cooling until its moisture condenses. What differs are the *causes* of the rising air: mountains for relief rainfall, air meeting along fronts in a depression for frontal rainfall, surface heating for convectional rainfall.

Frontal depressions

Q1 State temperature, cloud type and weather conditions at places A–D.

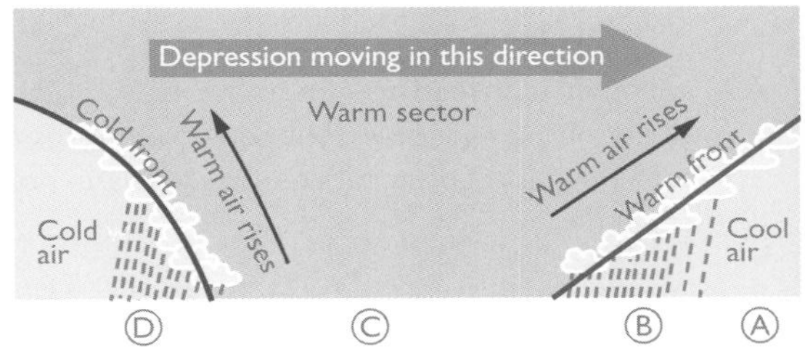

Q2 Explain the formation of frontal rain.

ANSWERS

A1 A = cool, high, thin cirrus cloud, dry
B = cool, lower, thicker stratus cloud, rain falling
C = warm, clear sky, dry
D = cold, tall cumulus/cumulonimbus clouds, heavy rain, possibly thunder

A2 warm air is forced to rise above colder air at the warm and cold fronts; rising air cools; moisture condenses to form clouds and rain; air is forced up more quickly at the cold front, causing heavier rain

examiner's note Frontal rainfall is the main type of rainfall in the UK. After a cold front has passed, typical weather is sunshine and showers.

Convectional rainfall

Q1 Describe the appearance of a cumulonimbus cloud.

Q2 State two ways in which cumulus clouds are different from other clouds.

Q3 Where and when does the great surface heating needed for convectional rain occur most frequently?

Q4 What makes convectional rainfall different from other types of precipitation?

ANSWERS

heavy rain formed by currents of hot air rising from heated surfaces

A1 very tall, white and fluffy, often an anvil shape at the top

A2 • they have greatest vertical difference between top and bottom of the cloud (i.e. they are the tallest)
• they may have an anvil shape due to spreading out at the top

A3 above land surfaces in the tropics all year; above land surfaces in temperate lands in summer; above tropical seas at the end of the summer

A4 it is heavy; torrential downpours are normal, often accompanied by thunder and lightning; hailstones sometimes replace rain droplets

***examiner's* note** Convectional rainfall is the main type of precipitation in the tropics.

Tropical cyclones (hurricanes)

Q1 Which places are most often affected by tropical cyclones?

Q2 Why do cyclones occur only at one time of year?

Q3 Loss of life is usually greater in developing countries (LEDCs) than in developed countries (MEDCs). Explain why.

Q4 Suggest two reasons why the value of property damaged varies greatly from one cyclone to another.

ANSWERS

the world's most violent storms, with strong winds and heavy rain, caused by very low pressure

A1 coastal areas in the tropics, particularly the Caribbean (hurricanes), Bangladesh (cyclones) and the Philippines (typhoons) (all three names describe the same event)

A2 very warm sea surfaces are needed, which are only present after the hot season

A3 LEDCs are less prepared, e.g. fewer shelters and emergency services; satellite-based weather forecasts give advance warning in MEDCs

A4
- it depends where a cyclone hits, e.g. a densely populated city compared to a farming area
- property values are higher in rich MEDCs like the USA

examiner's note Cyclones are major natural hazards that can cause total economic disaster to any poor tropical country taking a direct hit.

Natural hazards

Q1 This pie chart shows loss of life from natural hazards. Describe what it shows about the importance of climate-related hazards.

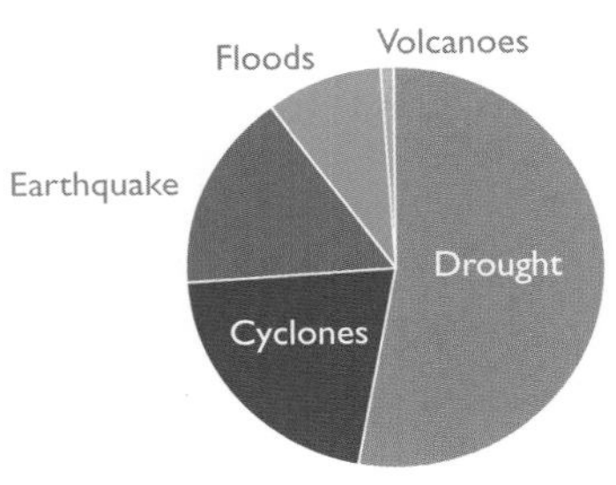

Q2 What is the difference between primary and secondary impacts of hazards on people?

ANSWERS

A1 drought, cyclones and floods are all climate related — the total percentage life loss from these is 83%; therefore they represent a great threat to people

A2
- primary impacts are the first or immediate impacts; people are killed and injured; buildings and crops are destroyed; emergency aid is usually needed
- secondary impacts are the effects in the short and medium term; shortage of food and water can lead to starvation and spread of disease; financial help for repairs and longer-term development aid are usually needed

examiner's note It is difficult to find any benefits or positive impacts of most natural hazards. Volcanoes (number 7) are the exception.

Ecosystems

Q1 Explain the two-way links shown in the diagram.

Q2 Why is climate the most important element in large ecosystems?

Q3 Name the two forest ecosystems of the British Isles.

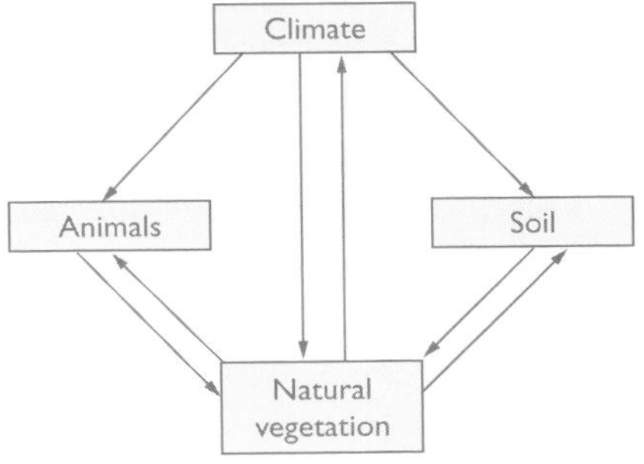

ANSWERS

natural systems in which physical and living elements are linked together

A1 temperature and precipitation affect type of vegetation; vegetation returns moisture to the atmosphere by transpiration; plants use soil nutrients, which are returned when they die (nutrient recycling); many animals feed directly on plants and their fruits; nutrients are returned to the soil in animal wastes

A2 climate controls natural vegetation (type and amount) and vegetation forms the base of food chains for all animal life; climate controls weathering, which breaks rock down into soil

A3 deciduous and coniferous

***examiner's* note** Because all elements in an ecosystem are interlinked, if one changes it has knock-on effects for all the others.

Tropical rainforests

Q1 Where are tropical rainforests located?

Q2 Describe the different layers that make up a rainforest.

Q3 'Rainforests contain the greatest biodiversity on Earth.' Explain what this sentence means.

Q4 Describe how the rainforest climate gives such good conditions for plant growth.

dense forest growing in hot, wet tropical lowlands near the equator

A1 lowland areas around the equator in South America, Africa and Asia; the largest remaining area is in the Amazon basin

A2 canopy cover of tall trees at about 30–35 m; above it, very tall individual trees (emergents); below it, layers of smaller trees; a layer of herbs and ferns on the forest floor

A3 there are more varieties of plants and animals in rainforests than in any other ecosystem

A4 the weather is hot (27°C) and wet (over 2000 mm rainfall) throughout the year, so the growing season is continuous

examiner's note Always use supporting values when referring to climate for both description and explanation.

Rainforest deforestation

Q1 Why are there strong economic pressures for developing countries (LEDCs) to cut down rainforests?

Q2 State the arguments of environmentalists against rainforest clearance.

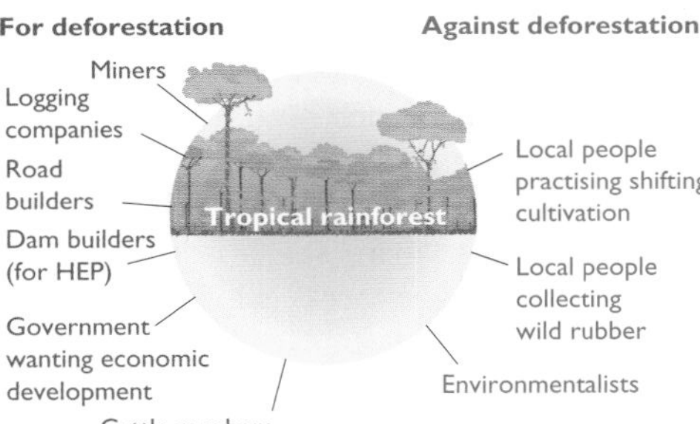

ANSWERS

A1 governments in LEDCs want more economic development; hardwood trees like mahogany and teak have great commercial value; valuable minerals lie below some forests; clearance gives new land for farming and settlements

A2 local consequences include increased runoff, soil erosion and flooding; the rich nutrient cycle is broken and soils soon lose their fertility; globally, forests are a great store of carbon dioxide; they maintain biodiversity of potential value for new crop seeds and medicines; many species of plants, animals and microorganisms are destroyed as forest is lost

examiner's note Governments in LEDCs argue that they need to make more and better use of their natural resources for their growing populations.

Sustainable development

Q1 How may reducing or stopping deforestation help each of the following global signs of human-caused unsustainability?

(a) global climate disruption
(b) changes in the water cycle
(c) spreading desertification
(d) soil erosion and land degradation
(e) loss of biodiversity
(f) wildlife habitats destroyed

Q2 State examples of sustainable ways of using natural forests.

A1 (a) forests are carbon sinks/stores, removing carbon dioxide from the atmosphere; (b) forests reduce runoff and increase evapotranspiration; (c) forests protect surfaces from wind and water erosion; (d) forests provide shelter and nutrients that improve soil structure; (e) forest ecosystems house the greatest variety of plants and animals; (f) forests are habitat providers and food producers

A2 selective logging of certain mature trees and re-planting; gathering and using natural forest products; creating National Parks and nature reserves to earn income from eco-tourism

examiner's note Sustainable development can never be fully achieved but conservation measures and better management of forest resources can help.

Desertification

Q1 One cause of desertification is the removal of surface vegetation. Give two reasons why trees and bushes are cut down by local people in developing countries (LEDCs).

Q2 Poor farming practices are a second cause. How can farming lead to bare surfaces and soil erosion?

Q3 Explain how climate may speed up the rate of desertification.

Q4 State the consequences of desertification for people in affected areas.

ANSWERS

the spread of desert-like surfaces into non-desert areas, mainly by the actions of people

A1 • to provide more farmland for rising populations
• for fuelwood, often the only energy source in rural areas

A2 overgrazing and over-cultivation; overuse of land destroys vegetation and soil structure; bare surfaces are exposed to erosion by water and wind

A3 if there are several drier than average years or if there is a period of drought

A4 lack of food leading to malnutrition and famine; death among the young and old in particular; forced migration

***examiner's* note** Desertification is a serious environmental and human problem, seen at its worst in the Sahel in Africa along the southern edge of the Sahara.

Pollution

Q1 What is the single main cause of air pollution?

Q2 Why is air pollution a serious problem in most big cities?

Q3 Name two effects of air pollution that are international (go across borders).

Q4 Name and describe two other types of pollution.

contamination of the natural environment by human activities

A1 burning fossil fuels

A2 higher volumes of traffic, traffic congestion, concentration of factories and houses; can be bad in LEDC cities where pollution controls are lacking or not enforced; cities like Los Angeles and Mexico City are well-known for air pollution — they are located in basins without much rain or strong winds to clear pollutants

A3 acid rain and global warming

A4
- land pollution, e.g. dumping plastics, which do not break down easily
- water pollution, e.g. chemicals in rivers, oil spills at sea
- other possibilities include noise and visual pollution

examiner's note Rising world populations and higher levels of economic development can only make pollution worse.

Acid rain

Q1 Name the two gases most responsible for acid rain.

Q2 From where are these gases emitted?

Q3 Describe the effects of acid rain.

Q4 State three methods of reducing levels of acid rain.

ANSWERS

rainwater with increased acidity due to air pollution

A1 sulphur dioxide and oxides of nitrogen

A2 coal-fired power stations; vehicle exhausts; factories

A3 trees lose leaves and die; crop yields reduced by more acid soils; plants and fish in lakes die due to acid water; limestone buildings and statues are dissolved away

A4 use limestone to absorb sulphur dioxide in power stations; fit catalytic converters to cars; use cleaner fossil fuels (e.g. natural gas) or alternatives (e.g. wind power) for electricity generation

***examiner's* note** Winds carry pollutants and acid rain from the UK to other countries, e.g. to Swedish forests and lakes, due to prevailing westerly winds.

The greenhouse effect

Q1 Name the four principal greenhouse gases.

Q2 Give the main sources for each gas.

Q3 Explain how the greenhouse effect operates.

Q4 How is its operation different from that of the hole in the ozone layer?

ANSWERS

heat energy from the Earth's surface is trapped by greenhouse gases in the atmosphere

A1 carbon dioxide, methane, CFCs and nitrogen oxides

A2 • carbon dioxide: burning fossil fuels and trees
- methane: decomposition of wastes, rice and cattle farming
- CFCs: fridges, aerosol sprays and air conditioning
- nitrogen oxides: car exhausts and chemical fertilisers

A3 sunlight heats the Earth's surface; greenhouse gases in the atmosphere trap heat in increasing amounts; less heat escapes into space; heat trapped near the surface leads to global warming

A4 less ozone is present to filter the sun's ultraviolet rays; surface temperatures are not affected but the risk of skin cancers is increased

***examiner's* note** Do not confuse the greenhouse effect and the hole in the ozone layer; they are unrelated.

Global warming

Q1 Describe what the graph shows.

Q2 Why are some places more worried about global warming than others?

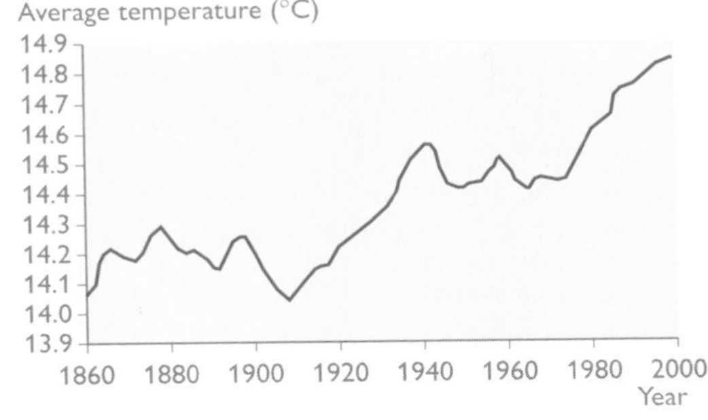

ANSWERS

the rise in average world temperatures

A1 an increase in average temperature of at least 0.7°C between 1860 and 2000; persistent increase despite falls, e.g. around 1910; steeper increase from 1980 onwards

A2 places with low-lying coastlines are most worried, e.g. delta countries like Bangladesh and coral islands like the Maldives, because they will be worst affected by rising sea levels; glaciers in the Alps will melt, affecting ski resorts; some countries might benefit, e.g. warmer climate in the UK for new crops and tourism, but no-one knows what the real effects of global warming might be

***examiner's* note** A rise in world temperatures is already observable. Is its cause natural, human or a mixture of both? This is an important point open to debate, but more and more people are blaming human activities.

Birth rate

- Average birth rate in developed countries (MEDCs): 12 per 1000
- Average birth rate in developing countries (LEDCs): 25 per 1000

Q1 State one economic reason for low birth rates in developed countries.

Q2 State one social reason for low birth rates in developed countries.

Q3 Give reasons why birth rates are high in many developing countries.

the number of live births per 1000 people per year

A1 people can afford to use birth control; children are expensive to raise; women wish to follow their own careers

A2 having one or two children is considered a normal family size; women are well educated

A3 reasons include lack of family planning clinics in rural areas; women are poorly educated and marry young; some governments and religions do not approve of birth control; children help with the family income

***examiner's* note** Economic reasons are related to money or wealth; social reasons are related to people, their traditions and their attitudes.

Death rate

- Average death rate in developed countries (MEDCs): 10 per 1000
- Average death rate in developing countries (LEDCs): 9 per 1000

Q1 Explain why death rates have declined almost everywhere in the world during the last 50 years.

Q2 Why are death rates in developed countries similar to those in developing countries despite the better medical facilities developed countries have?

Q3 Name a country with an increasing death rate.
Give a reason for this increase.

A1 due to the spread of improved medical knowledge and health-care; vaccination programmes against killer diseases; improved access to clean water supplies

A2 in developed countries there are more old people who are reaching the end of their natural life spans whereas a higher percentage of young people live in developing countries

A3
- some African countries such as Zimbabwe, Botswana, Sierra Leone or Rwanda; also Russia
- people in many southern African countries are badly affected by AIDS; in other countries it is due to wars; in Russia due to poverty

***examiner's* note** Because death rates are low almost everywhere, differences in birth rate have a greater effect on rates of population increase.

Demographic transition model (DTM)

Q1 What are the similarities and differences between the lines for birth and death rates?

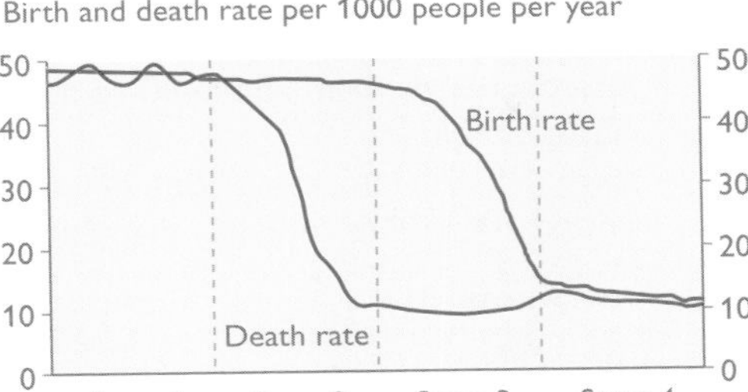

Q2 Describe and explain the differences between stages 2 and 4.

ANSWERS

line graph showing the relationship between birth and death rates as a country develops

A1 both lines start high, fall quickly and level off at a lower rate; the line for birth rate falls quickly in stage 3 compared with stage 2 for death rate; birth rate is higher than death rate for most of the time

A2
- stage 2: high birth rate (above 40) remains steady, but the death rate decreases from above 40 to about 10 due to medical improvements
- stage 4: low birth rate (about 12) and death rate (about 10), almost stable due to family planning and healthcare

examiner's note The larger the gap between birth and death rate lines, the greater is the population growth. Some people now add a stage 5 to the DTM for countries with a higher death rate than birth rate, such as Italy (number 54).

53 ANSWERS

Natural increase and natural decrease

	Birth rate (per 1000)	Death rate (per 1000)
Ghana	39.2	11.2
UK	11.0	10.2
Italy	8.8	10.6

Q1 What is the rate of natural increase in Ghana and the UK?

Q2 Calculate the rate of natural decrease in Italy.

Q3 Give the number of the stage reached in the demographic transition model by Ghana and the UK. Explain your choices.

ANSWERS

the difference between birth and death rates

A1 28 (39.2 – 11.2) and 0.8 (11.0 – 10.2) per 1000 people

A2 – 1.8 (8.8 – 10.6)

A3
- Ghana: stage 2; UK: stage 4
- big difference between high birth rate and low death rate in Ghana, causing a high natural increase
- in the UK, both birth and death rates are low

examiner's note Italy has moved from stage 4 to stage 5 in the DTM because its death rate is now higher than its birth rate to give it a natural decrease.

Population structure

Q1 Describe the shape of this population pyramid.

Q2 How does it show a young population structure?

Q3 Name some problems caused by young population structures.

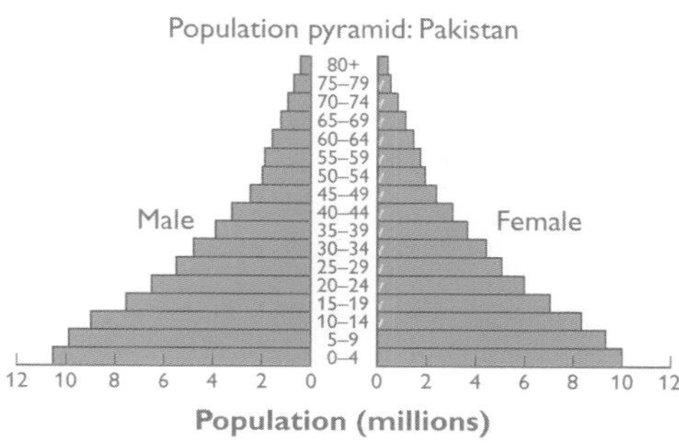

the make-up of a population by age and sex, usually shown by a population pyramid

A1 wide base, steep sides and narrow top; almost a perfect pyramid

A2 widest at the base; almost 60 million people below the age of 14; much lower numbers aged 45 and above

A3 high population increase; increased demands for food; pressure on services such as education and health; growth of big cities and associated urban problems; unemployment and poverty; environmental damage (soil erosion, water and air pollution)

***examiner's* note** Pakistan's population pyramid is typical of that for many developing countries. Look first at the bottom of population pyramids because this shows what has happened most recently.

Ageing population

Q1 Describe the shape of the population pyramid.

Q2 How does the pyramid show an ageing population?

Q3 Name two costs and two benefits of an ageing population.

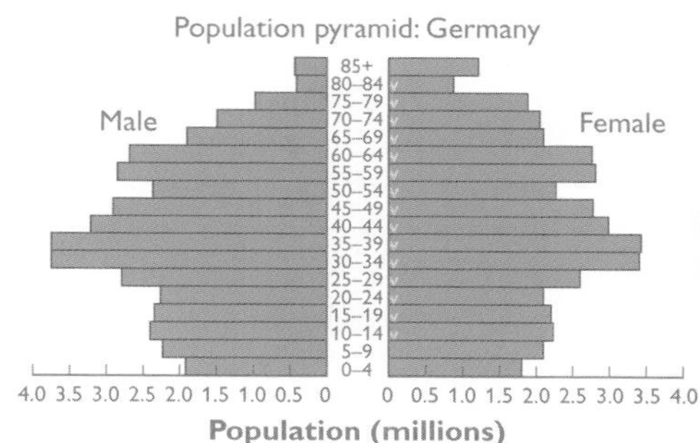

the increasing percentage of old people (aged 65 and over) in a country

A1 widest in the middle (e.g. 30–44); narrower at the base; irregular shape

A2 large numbers of people aged over 65; quite wide at the top, especially females; about 3 million people aged over 80

A3
- costs: government pays out more on pensions; health and care services for the elderly are expensive; working population pays for these through taxes
- benefits: elderly people provide a market for off-peak holidays and use of leisure facilities mid-week; companies profit by making products for the elderly such as stair lifts

***examiner's* note** This is an issue of increasing importance in MEDCs with a natural decrease (e.g. Germany's birth rate is 8.7, death rate 10.7: a decrease of 2.0 per 1000).

Migration

Q1 State the differences between the pairs of migration terms given in (a), (b) and (c).

(a) forced voluntary
(b) national international
(c) permanent temporary

Q2 What is meant by the terms 'refugee' and 'economic migrant'?

the movement of people from one place to another to live

A1 A = people are driven out in forced migration whereas they choose to move in voluntary migration; B = 'national' is migration within a country whereas 'international' is moving to another country; C = 'permanent' is staying for ever in a new area or country whereas 'temporary' means going back home at a later date

A2
- a refugee is a person forced to flee from the country where they live due to natural disasters (e.g. flood, drought, volcanic eruption) or human factors (e.g. war)
- an economic migrant is someone who moves for work, such as Mexicans migrating to the USA

examiner's note People migrate due to push and/or pull factors.

Rural-to-urban migration

Q1 Identify and name the push factors shown in the diagram.

Q2 Which factor is a physical push factor?

Q3 Describe some of the pull factors of urban areas in developing countries (LEDCs).

the movement of people from the countryside into cities, most often in developing countries

A1 shortage of land; insufficient food; drought; soil erosion; lack of services (e.g. no electricity)

A2 drought (a climatic factor caused by lack of rainfall); soil erosion is a possible answer, but it is less effective because human factors (i.e. poor farming methods) are its main cause

A3
- jobs: greater number; more variety; better paid
- more services: electricity; schools (including secondary schools); hospitals
- higher standards of living with more modern facilities

***examiner's* note** The unstoppable movement of people into cities is causing massive urban problems (poor housing, inadequate services, traffic congestion and water and air pollution). Cities are more dynamic than the countryside.

Urban-to-rural migration

Q1 Name three problems for people living in large urban areas in developed countries (MEDCs).

Q2 Give two examples of urban decay in British cities.

Q3 State some of the attractions of living in a small rural village in the UK.

Q4 Why do some local people object to city people moving into their villages?

ANSWERS

the movement of people from cities into the countryside, most often in developed countries

A1 poor-quality housing; traffic congestion; lack of open spaces; pollution (fumes from traffic, noise pollution from transport and visual pollution from old industries and docklands)

A2 old, badly maintained terraced houses; derelict land such as old railway sidings, abandoned factories and warehouses

A3 peace and quiet; clean air; good appearance; close to open countryside

A4 house prices increase; shortage of houses for local people; no support for village services (e.g. commuters do not use local shops); traffic increases on country roads; pressure for new growth

examiner's note Villages close to large cities and in areas of attractive countryside are most at risk of urban-to-rural migration.

Site, situation and layout of settlements

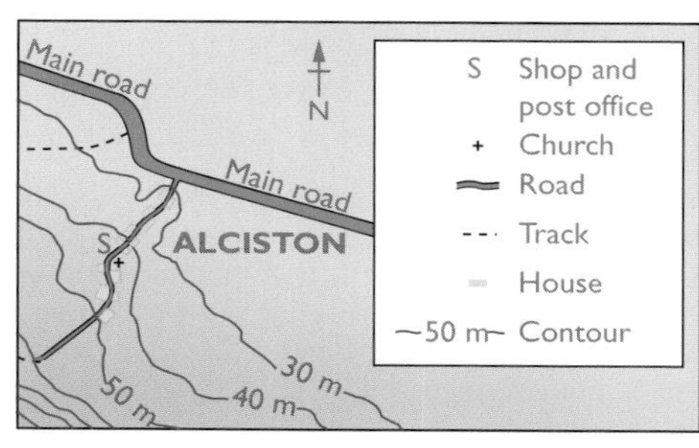

Q1 The diagram shows the village of Alciston in Sussex. Describe its site.

Q2 Give its situation.

Q3 State its layout.

ANSWERS

A1 on gently sloping land, 30–50 m above sea level

A2 on a side road close to the main road, on sloping land between the steep (scarp) slope to the south and the flat lowland to the north

A3 long and narrow along the road side (linear or ribbon shaped), without any housing developments away from the road sides

examiner's note Site refers only to the land on which the settlement is built whereas situation is a broader term referring to position in relation to the surrounding area. Layout is the shape or arrangement of the settlement, or its form. Other common forms for small settlements are clustered around a road junction or bridge and oblong or square around an open space in the centre (such as a village green).

Hierarchy of settlements

Q1 Name the *rural* settlements shown in the diagram.

Q2 Describe what changes in settlement size, number of services and sphere of influence occur going up the hierarchy.

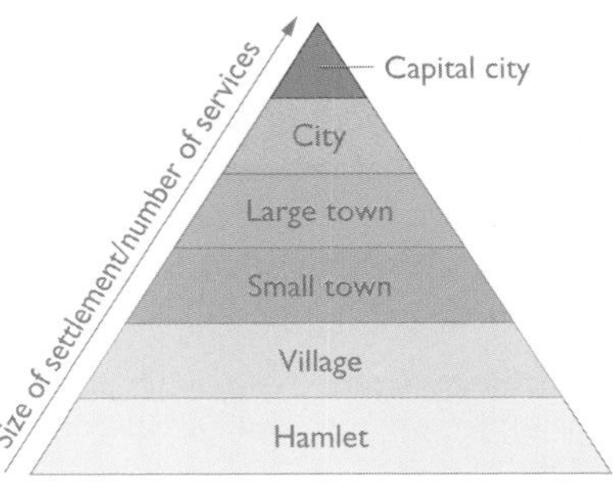

ANSWERS

the placement of settlements in order of size and importance

A1 hamlet and village, perhaps a small town with strong rural links

A2
- settlement size: increases from only a few people in a hamlet to possibly millions in a big city
- number of services: increases from none in a hamlet, to perhaps only a shop, pub and church in a village, to hundreds of shops and services in a large town; cities and capital cities have additional services such as large hospitals and airports
- sphere of influence: increases from serving the local area only (rural settlements) to drawing people from a large area to shop (urban settlements); for a capital city it is the whole country

***examiner's* note** A hierarchy is always a pyramid shape because numbers of small settlements are always greater than those of large settlements.

Urban morphology

Q1 State the shape of the urban model.

Q2 Describe and explain changes in land use from the centre to the edge of the city.

Burgess model

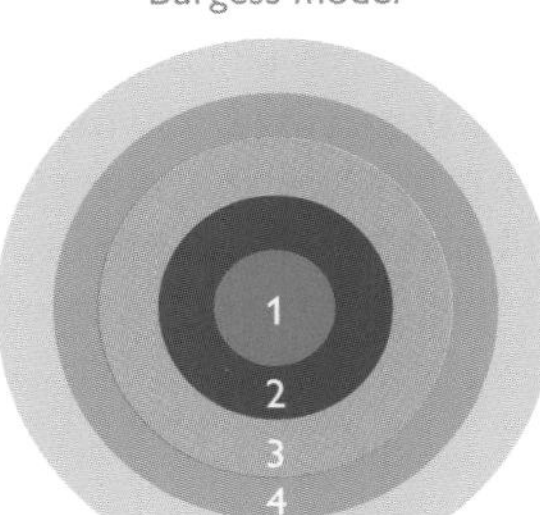

1 Central business district (CBD)
2 Inner city
3 Older residential suburbs
4 Newer residential suburbs
5 Rural–urban fringe

ANSWERS

the shape (form) and structure of towns and cities

A1 circular urban zones around a central CBD

A2
- land use changes from being dominated by business in the centre to residential housing in the suburbs; then it becomes less built-up in the rural–urban fringe
- cities grow from the centre outwards; businesses such as shops and offices became more concentrated in the centre; the centre is surrounded by places for people to live, with the newest housing usually near or beyond the city edge

***examiner's* note** The circular pattern of urban zones around the CBD in the centre can be recognised in most cities, even if only in part.

Urban zones: the central business district (CBD)

Q1 Describe the main characteristics of the CBD.

Q2 Why is it often easy to spot the CBD on a photo of a city?

Q3 Name two land uses that cover larger areas in other urban zones than they do in the CBD.

Q4 Explain why these land uses are less important in the CBD.

the urban zone in the city centre dominated by shops and housing

A1 the largest concentration of offices and shops including department stores, with the widest variety of goods for sale; the main place of work by day; rush-hour traffic congestion; high rents and rates; the point where main roads meet; the most densely built-up area

A2 the concentration of skyscrapers and other tall buildings

A3 housing, industries and open spaces

A4 high demand for land makes it too expensive for these land uses; rents and rates are too high for them

examiner's **note** The CBD is the most dynamic zone, with high levels of business activity and rapid rates of change.

Urban zones: inner city

Q1 State some of the land uses found in the inner-city zone.

Q2 Explain why many inner-city zones are places of urban decay.

Q3 Name an example of a redeveloped inner-city area.

Q4 Describe some of the changes made during the redevelopment of inner-city areas.

the old urban zone next to and surrounding the CBD

A1 old houses (mainly terraced); factories and warehouses; derelict/waste land (e.g. railway sidings, old docks); tower blocks of flats

A2 many of the houses, factories, warehouses and railways were built more than 100 years ago when industry was more important; modern businesses and builders of new houses prefer locations nearer city edges

A3 Docklands in London; Albert Dock in Liverpool; or a local example

A4 warehouses converted into luxury flats; docks changed into marinas; waste land reclaimed for houses and businesses; this is often called gentrification

examiner's note The inner city is the most mixed zone in UK cities, with old and new, smart and ugly often found next to each other.

Brownfield sites

Q1 State three land uses shown in the photo.

Q2 What shows that this is a brownfield site?

Q3 Why is it not a greenfield site?

ANSWERS

areas of previously built-up land that can now be reused for building

A1 blocks of flats; gas holders; old warehouses/factory buildings; equipment storage; weed-covered waste land

A2 it is likely that land in the foreground has been cleared of buildings and abandoned as derelict; old factory/warehouse partly bricked up suggests that it is no longer used and could be knocked down and land cleared for new building

A3 in wrong location (not rural); inner-city land like this must have been built on before; greenery is from weeds, not farmland in the countryside

examiner's note It is more expensive to clear and build on brownfield than on greenfield sites, but improvements to the environment are more likely when brownfield sites are used.

Urbanisation

Q1 This graph shows total world urban population. Describe the changes it shows.

Q2 Give two reasons for high rates of urbanisation in LEDCs.

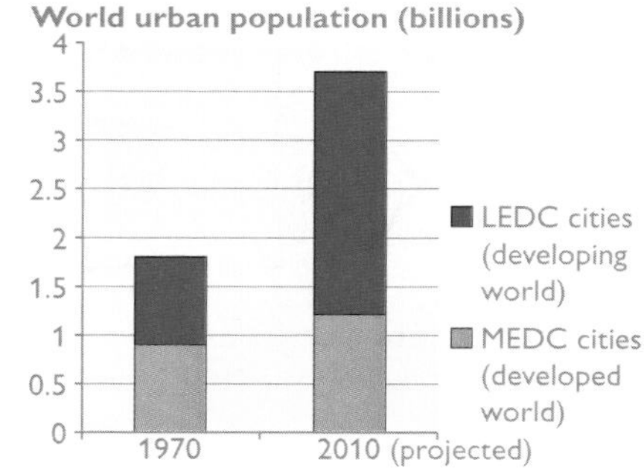

ANSWERS

an increase in the percentage of people living in urban areas

A1 total urban population doubles between 1970 and 2010; most of the increase is in LEDCs; the percentage living in LEDC cities increases from one half to two thirds of total urban population

A2
- high rates of rural–urban migration due to push–pull factors, with more services and more varied work available in the big cities than in the countryside
- high rates of natural increase of population due to less use of birth control and young population with many people of child-bearing age

***examiner's* note** In MEDCs, high percentages of people already live in cities so that increases in urban population today are greatest in big cities in LEDCs.

Shanty towns

Q1 Describe the housing shown in the photo.

Q2 Where is this shanty town located?

Q3 State the urban problems in shanty towns.

areas of low-cost, poorly built housing without most or all essential services

A1 makeshift housing built by residents; wooden frames; sheets and cloth of many different sizes and colours used; some metal sheet roofs; fronts overhang onto street; houses crammed together

A2 next to a railway line

A3 no waste collection/a lot of rubbish; shortage of supplies of clean drinking water; absence of a sewerage system; frequent spread of disease; lack of other services such as electricity; overcrowding; crime; difficult for shanty dwellers to find paid work and break out of the poverty trap

***examiner's* note** In most big cities in LEDCs, one third or more of inhabitants live in shanty towns because city authorities cannot afford to build formal housing.

Urban problems

List the following urban problems under the headings 'socioeconomic' and 'environmental'.

- air pollution
- lack of essential services
- polluted rivers and waterways
- traffic congestion
- urban sprawl
- crime
- noise
- slum housing
- unemployment

Why are urban problems difficult to solve?

ANSWERS

exist in all cities due to the great concentration of people and economic activities

A1
- socioeconomic: crime, lack of essential services, noise, slum housing, unemployment
- environmental: air pollution, polluted rivers and waterways, traffic congestion, urban sprawl

A2 fast and continuous city growth, no time to catch up with existing problems; expensive to solve, e.g. building houses, providing clean water, sewerage and electricity; cars and traffic increase with economic growth and development; economic growth is usually considered more important than concerns about the environment

***examiner's* note** Urban problems are not as severe in cities in MEDCs because more money is available and because pollution regulations are enforced.

Sustainable urban living

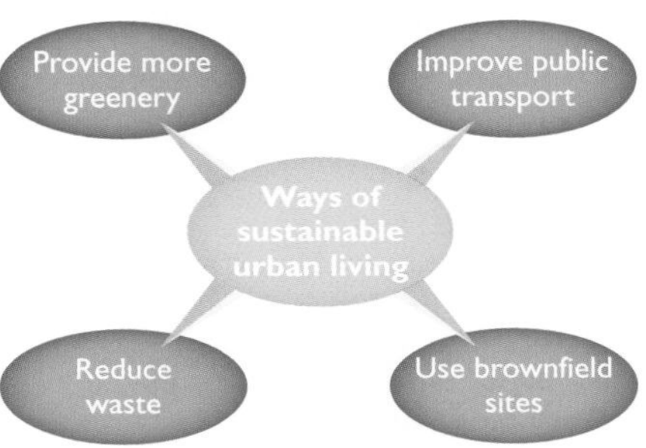

Q1 Give examples of each of the four ways for sustainable urban living shown in the diagram.

Q2 What is an eco-town?

ANSWERS

improving quality of living in towns and cities while reducing damage to the environment

A1 • provide more greenery: plan for areas of open green spaces; landscape the sides of paths, waterways and railways
• improve public transport: link bus, rail and tram routes at transport interchanges; use more environmentally friendly buses
• use brownfield sites: use waste or empty land instead of greenfield sites in the countryside; landscape and improve them
• reduce waste: reuse bottles, plastic bags etc. more than once; recycle glass, paper, plastics, cans; up to 90% of household waste is recyclable

A2 settlements with low energy consumption, carbon neutral or zero carbon emissions, built from recyclable materials, green spaces up to 40% of the area

examiner's note Most urban living is not sustainable at present. New initiatives aim to make big changes to the way people live.

Farming terms and types

Q1 What is meant by the terms 'arable', 'pastoral' and 'mixed' farming?

Q2 State the three key differences between intensive and extensive farming.

Q3 What is a subsistence farmer?

Q4 In what ways does commercial farming differ from subsistence farming?

ANSWERS

include arable and pastoral, intensive and extensive, subsistence and commercial

A1 • arable: growing crops
- pastoral: keeping animals
- mixed: growing crops and keeping animals

A2 intensive farming gives a higher output; more inputs are used on an intensive farm; a larger area of land is needed in extensive farming

A3 someone who grows crops and keeps animals to feed their family (with little or none for sale), most commonly found in the tropics

A4 in commercial farming crops are grown and animals reared for sale; farming is a business; money made from selling farm produce is reinvested in the farm to increase the produce for sale

***examiner's* note** All types of farming can be described using these terms.

The farm system

Q1 Name the physical inputs. Why are they important?

Q2 Describe the processes on a pastoral farm.

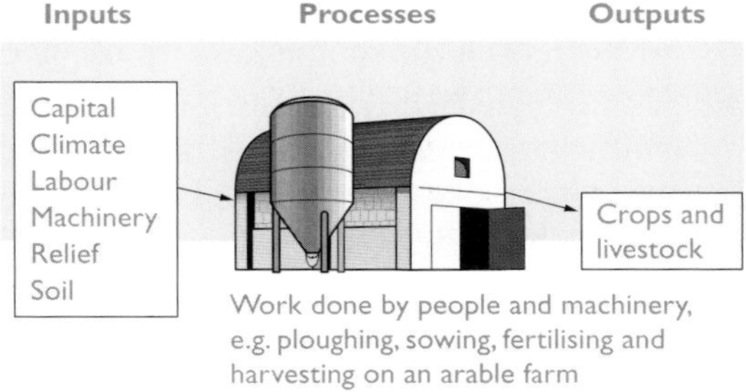

Q3 Name an output that can be reused on a farm.

Inputs	Processes	Outputs
Capital Climate Labour Machinery Relief Soil	Work done by people and machinery, e.g. ploughing, sowing, fertilising and harvesting on an arable farm	Crops and livestock

ANSWERS

A1 • climate, relief and soil
• they are important because warm temperatures, water and fertile soils are needed to make crops and grass grow well; if the relief is gentle it is easier to use machinery

A2 milking cows; feeding animals; turning cattle out into the fields; moving them to new pastures; lambing; dipping; shearing and rounding up of sheep with dogs; cutting hay and making silage

A3 animal manure spread on land as fertiliser; crops kept as winter feed for animals

***examiner's* note** The greater the human inputs into the system, the greater the outputs are likely to be, which is why intensive farming gives the greatest output.

The Green Revolution

Q1 Name the main crops grown using high-yielding seed varieties.

Q2 Give an example of a new variety of seeds.

Q3 State the advantages for LEDCs of using the new seeds.

Q4 Explain two disadvantages of the Green Revolution.

the increase in food production due to the use of new, high-yielding varieties of seeds

A1 maize (corn), rice and wheat

A2 IR-8, a variety of rice grown in the Philippines, which increases yields six-fold

A3 provides extra food for growing populations; helps to prevent famine; countries need to import less food; improved living standards in rural areas

A4
- often the high yields are achieved only by using more fertiliser and irrigation water, which many poor farmers cannot afford
- big farmers become richer and buy up more land so that the wealth gap between rich and poor farmers widens

***examiner's* note** The Green Revolution has been a success in Asia, a continent with large farming populations.

Environmental problems from farming

Q1 Why did some British farmers remove hedgerows?

Q2 State the environmental problems that resulted.

Q3 Why were nitrates and phosphates reaching rivers in increasing amounts?

Q4 Describe the environmental problems caused by them.

ANSWERS

these are increasing as land is used more intensively to raise output

A1 for easier use of larger machinery; to increase the amount of cropland

A2 loss of wildlife habitats; decrease in animal and bird life; more soil erosion from strong winds due to less shelter

A3 both are common ingredients in fertilisers, which are used to increase crop yields

A4 water is enriched, encouraging the rapid growth of algae, which uses up oxygen; there is insufficient oxygen left in the water for other plant and animal life, which eventually dies; this process is called eutrophication

***examiner's* note** Problems are greatest in crop-growing areas like East Anglia, where pesticides that kill off harmless species are also used.

The CAP (Common Agricultural Policy)

Q1 What was the main aim of setting up the CAP?

Q2 What were the policy's major weaknesses?

Q3 Why was the Environmental Stewardship Scheme introduced in 2005?

Q4 How does it work?

A1 to make the EU self-sufficient in food production

A2 over-production leading to grain mountains and wine lakes; high food prices; large payments to already rich big landowners; unfair trade for developing countries

A3 to offer incentives to improve landscape quality and conserve wildlife

A4 farmers are paid to manage the countryside, making new hedgerows, woodlands and wetlands — useful as wildlife habitats

***examiner's* note** The CAP badly needed reform. Emphasis on the environment is increasing at the expense of farm output

Diversification in farming

Q1 In which of the following ways are farmers (a) living close to a city; (b) in a National Park most likely to diversify? Explain why.

- farm shops
- making ice cream and yoghurt
- cafés and tearooms
- camping and caravan sites
- PYO (pick your own)
- forestry
- bed and breakfast
- horse riding

Q2 What type of farmer is most likely to start making ice cream and yoghurt? Why can this be more profitable than selling farm produce to a processor or distributor?

ANSWERS

starting new money-making activities to broaden the sources of income away from farming

A1 (a) close to a city: farm shops, PYO and horse riding are most suitable for earning money from the nearby urban dwellers

(b) in a National Park: cafés and tearooms, bed and breakfast, and camping and caravan sites are suitable for tourists in National Parks

A2 a dairy farmer; by making a product the farmer is adding value and earning more money from each litre of milk compared with selling it to a dairy

***examiner's* note** Opportunities for diversification are not the same everywhere, with less choice for farmers remote from towns and tourist areas.

Sustainable rural living

Q1 Give examples of each of the four ways for sustainable rural living shown in the diagram.

Q2 What are the problems with achieving these?

ANSWERS

improving quality of living in the countryside, and the natural landscape and environment

A1
- protect the countryside: preserve green belts; conserve areas with National Parks, nature reserves and Areas of Special Scientific Interest (SSSIs); Environmental Stewardship Scheme for farmers
- increase accessibility to services: extend broadband internet to reduce isolation and allow doing business from home
- diversity into non-farming activities: into tourism; recreation; making foodstuffs
- affordable homes: Home Starter Initiative; homes set aside for local people only; full council tax charged on second homes

A2 services are more expensive to provide in rural areas; bus services, shops and pubs are decreasing; the countryside attracts urban workers with higher wages

examiner's note Economic changes are out of the government's control

Irrigation

Q1 Describe the irrigation method shown in the diagram.

Q2 State two advantages of this method.

Q3 Explain where irrigation is used.

artificially adding water to farmland to make crops grow

A1 water, carried in a pipe, is released around the plants

A2 • less water is wasted and lost to evaporation
• there is less danger of salinisation (soil becoming salty) because crops are using water before it can be evaporated and leave behind salt

A3 desert areas with little rainfall, e.g. Nile valley in Egypt; places where rain falls only in one season and it is dry for the rest of the year, e.g. the Ganges valley in India; in wet climates to increase yields, e.g. potato farmers in East Anglia

***examiner's* note** Most irrigation water is stored behind dams and released when needed.

The factory system

Q1 Name and define at least four inputs into the factory system.

Q2 Why is waste an output from the factory system?

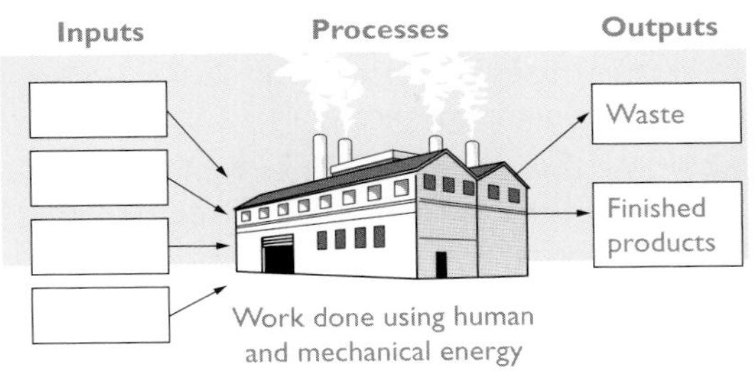

ANSWERS

A1
- raw materials: natural resources used for making a new product
- energy/power: fuel or electricity to drive the machinery
- capital: money invested into the business
- labour: people to operate the machines and provide special skills
- transport: movement of raw materials and fuels by road, rail or water
- government grants: aid for setting up factories, e.g. in areas of high unemployment

A2 waste (mainly gases) comes from burning fuels and from those parts of the raw materials not used for the final product (mainly solid waste)

***examiner's* note** The relative importance of the inputs varies with the type of manufacturing industry. Raw materials and energy are important for heavy industries, whereas skilled labour is much more important for high-tech industries.

Types of manufacturing industry

Q1 Name two examples of heavy industries.

Q2 What are the main characteristics of heavy industries?

Q3 Why are many light industries also known as consumer industries?

Q4 Name two types of high-tech industries. Why are these examples of 'footloose' industries?

these range from heavy to light, including the newer high-tech industries

A1 steel making, shipbuilding, oil refining and petrochemicals

A2 they make large or bulky products; they use large amounts of raw materials and fuel; they have huge works covering large areas of land

A3 they produce goods for people to buy such as electrical goods, clothing, food and drink

A4
- computers, telecommunications equipment (e.g. mobile phones), microelectronics
- they have great freedom in choice of location because they do not need bulky raw materials; their factories can be located anywhere

***examiner's* note** Each type of industry has different location needs; high-tech industries are increasing fastest due to economic development and advances in technology.

Location of light industries in the UK

Q1 Describe the transport advantages of the M4 corridor for light industries.

Q2 Explain other favourable factors for the location of light industries here.

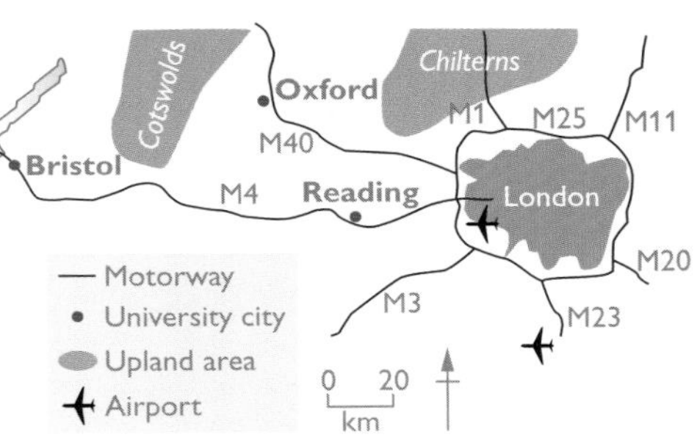

depends most on easy access to markets because fuel and raw material needs are low

A1 the M4 is a motorway giving fast links between London and Bristol; it links to the M25 from which motorways go to the rest of the UK; it passes close to Heathrow airport

A2 London is the largest city and largest market in the UK; university cities supply graduates and skilled workers for research work into new products; upland areas nearby provide pleasant countryside for recreation at weekends; proximity to London, Heathrow airport and countryside makes it easier to attract quality workers

***examiner's* note** Growth industries prefer locations close to large centres of population and good transport links.

Old industrial locations in UK cities

Q1 What features show that this is an old factory in an inner-city location?

Q2 Describe the disadvantages of inner cities today for factory locations.

mainly in the inner-city zone just outside the CBD

A1 the style of brickwork, type of building and the chimneys suggest that it is an old building; it is located next to a canal, which was more important for transport in the past than today

A2 old, often narrow, roads and streets congested with traffic, unsuitable for today's large delivery lorries; absence of parking spaces for workers' cars; shortage of space for expansion; poor environmental quality of surrounding areas; workers more likely to live in the suburbs than close by

***examiner's* note** Some industries stay in the same location because it is too expensive to move. This is known as industrial inertia.

New industrial locations in UK cities

Q1 What is an industrial estate?

Q2 A few industrial estates are found in inner-city locations. Explain why.

Q3 Most are located on the edges of cities. Give the advantages of an edge-of-city location for industries.

Q4 Is an edge-of-town location a brownfield or a greenfield site? Explain your answer.

mostly on planned industrial estates, usually located on the edges of cities

A1 an area laid out for factories and businesses, provided with essential services like electricity and roads

A2 they occupy waste land; can be part of an inner-city improvement scheme; make use of a brownfield site and save more countryside

A3 nearer to motorways and A-roads without the need to pass through the traffic-congested city; more space for the factory buildings and car parks; nearer to workers' homes in the suburbs; occupy a more pleasant environment next to open countryside

A4 greenfield site; first-time building on land that was previously countryside (i.e. green fields)

***examiner's* note** Industrial estates are the preferred locations for most industries today.

Business parks

Q1 Describe three characteristics of the business park shown in the photo.

Q2 Where and why are most business parks located?

ANSWERS

areas laid out for offices, usually in out-of-town locations

A1 • large and modern buildings
- car-parking space
- landscaped areas
- empty spaces around the buildings
- offices for call centre

A2 • in out-of-town locations, particularly near junctions with access to motorways or A-roads; on greenfield sites
- easy access for workers and visitors; enough space to create a pleasant working environment; in locations close to open countryside

***examiner's* note** Business parks are a modern type of industrial estate and include warehouses for distribution and small factories as well as offices; call centres commonly locate on them. They are similar to the science parks for high-tech companies engaged in research.

Transnational corporations (TNCs)

Q1 Name an example of a transnational corporation with a business in each of the following: food and drink; oil and gas; motor vehicles.

Q2 Where do most transnationals have their head offices?

Q3 Many transnationals have factories in LEDCs. State two advantages and two disadvantages for LEDCs of the presence of transnationals in their countries.

large companies with business interests in many different countries (multinational companies)

A1
- food and drink: McDonald's, Coca Cola, Nestlé
- oil and gas: BP, Exxon, Shell
- motor vehicles: Toyota, Ford, VW

A2 in MEDCs — mainly North America, Europe and Japan

A3
- advantages: create jobs; encourage improvements in the infrastructure such as roads and energy supplies; set up industries that the countries could not do for themselves; increase exports
- disadvantages: pay low wages; profits made are taken out of the country; have poor safety records; the company rather than the government chooses which goods to make and export

***examiner's* note** Asian countries such as China, Malaysia and Thailand attract more TNCs than other LEDCs. African countries attract many fewer — why?

Newly industrialising countries (NICs)

Q1 Name three examples of newly industrialising countries.

Q2 What is the main reason for transnational corporations setting up factories in NICs?

Q3 State the other attractions that NICs have for transnational corporations?

Q4 Explain how the growth of manufacturing industry encourages economic development.

LEDCs, mainly in Asia and Latin America, experiencing a rapid growth in manufacturing industry

A1 Brazil, Mexico, Taiwan, Malaysia, Thailand, Indonesia and South Korea

A2 cheap wages

A3 plentiful and willing workforces, eager to work for overseas companies; growing home markets; good support from governments keen to attract new industries and different types of work

A4 manufacturing leads to higher incomes; workers are trained and learn new skills; education and public services improve; people's spending power increases, which supports other manufacturing industries and the growth of services

examiner's note Some former NICs such as Singapore have developed to the point where they have many of the characteristics of developed countries.

Industrial growth in East Asia

Q1 Describe and explain what the graph shows about the growth of industry in East Asia.

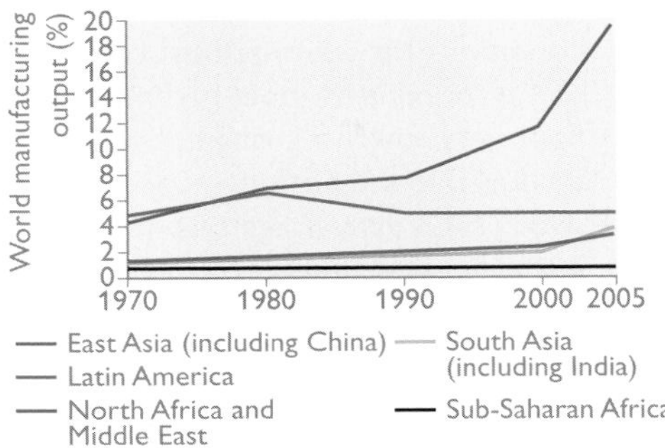

Q2 How and why is sub-Saharan Africa different?

A1 • describe: rising since 1970 when it was 4%; faster growth since 1990; growth has speeded up since 2000; by 2005 almost 20% of world manufacturing output in East Asia
• explain: cheap but reliable workforce; attractive political conditions for transnational corporations (TNCs); growth of China explains rapid recent growth

A2 • how: very low; no change; about 1% of world manufacturing throughout the period
• why: TNCs discouraged from setting up factories by political instability, wars and corruption, low levels of education and skills, poverty and small home market

***examiner's* note** Choose one sub-Saharan country; find out more about it to explain why it has little manufacturing industry (e.g. DR Congo or Zimbabwe).

Primary, secondary and tertiary employment

Q1 Name three types of work in the primary sector of employment.

Q2 What do they have in common?

Q3 How is work in the secondary sector different?

Q4 Give examples of different types of work in the tertiary (service) sector.

the three groups into which all types of jobs can be fitted

A1 farming, fishing, mining and forestry

A2 products from the land or sea are obtained; raw materials to eat or use are collected (but not made into other products)

A3 this is work in factories where raw materials are manufactured into other products

A4 providing public services (e.g. healthcare, council services, electricity); transport (buses, trains, planes); retail (shops, garages, supermarkets); financial services (banks, building societies); leisure and tourism (hotels, cafés, sports centres)

examiner's note The service sector is the fastest-growing sector in all countries.

Employment differences between MEDCs and LEDCs

Q1 Describe the differences in employment between MEDCs and LEDCs.

Q2 Give reasons for the differences.

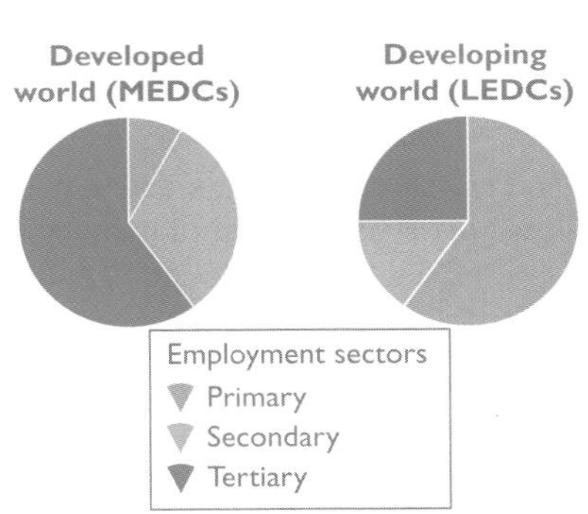

the gap in economic development, shown by the dominance of primary employment in LEDCs

A1
- largest sector in LEDCs is primary (about 60%) whereas it is the smallest in MEDCs (under 10%)
- largest sector in MEDCs is tertiary (about 60%) whereas it is much lower in LEDCs (25%)
- secondary sector in MEDCs is double the size of LEDCs

A2
- in LEDCs, most people are farmers; they are too poor to support a wide range of manufacturing and service industries
- in MEDCs, farming is mechanised; wealthy city dwellers can afford to pay for large numbers of goods and services

examiner's note Quote values when describing from graphs and use 'whereas' to highlight differences.

Advantages of tourism

Q1 Explain how tourism increases jobs in the service sector.

Q2 What is the main economic advantage of overseas visitors for a country?

Q3 State one environmental advantage of tourism.

Q4 'Tourism often leads to an improvement in infrastructure.' What does this mean?

mainly economic but can also be environmental and social

A1 tourist services are labour-intensive, with many workers employed in hotels, cafés and restaurants, in transport, supervising beach and other leisure activities, acting as tourist guides or working in souvenir shops

A2 it earns foreign exchange, used for buying goods from other countries

A3 encourages conservation of scenic areas and wildlife habitats; entrance fees help to cover costs of conservation and management

A4 infrastructure needed by tourists includes airports, roads, water, sewerage and electricity; tourism encourages improvements in standards

***examiner's* note** Tourism is one of the world's great growth industries and is a vital source of income and employment for some developing countries.

Disadvantages of tourism

Q1 List four economic disadvantages of tourism.

Q2 List two social disadvantages of tourism.

Q3 List three environmental disadvantages of tourism.

Q4 Explain what can be done to reduce these environmental disadvantages.

impacts of tourism, which tend to be more severe in LEDCs

A1 farmers lose livelihoods as tourist developments take up farmland; many jobs are seasonal; visitor numbers fluctuate; work in tourism is low paid

A2 conflicts between farmers and visitors; local traditions are lost

A3 footpath erosion; litter and sea pollution; sprawl of hotels along coast

A4
- footpath erosion: close sections off to allow recovery; repair surfaces; use flat stones and steps on badly eroded sections
- litter and sea pollution: clean-up patrols; sewerage systems to prevent untreated waste going into the sea
- sprawl of hotels along the coastline: plan hotel development; include measures to landscape the area; restrict size of developments

***examiner's* note** Good management of a country's tourist industry can greatly reduce the impact of these disadvantages.

Ecotourism ('green tourism')

Q1 'Ecotourism is environmentally sound tourism.' What does this mean?

Q2 'Ecotourism is socially sound tourism.' What does this mean?

Q3 Why is ecotourism more sustainable than mass tourism?

Q4 Where is ecotourism already being practised?

ANSWERS

tourism that is concerned with protecting the environment and way of life of the local people

A1 care is taken to safeguard natural environments and wildlife, protecting them from development and pollution

A2 care is taken to safeguard the traditional way of life of local people, to involve them in decisions and to give them a share in financial benefits

A3 tourism with a low impact on the local area and its people will be a viable long-term activity

A4 many rainforest lodges; countries rich in wildlife such as Costa Rica; wilderness areas such as Antarctica where regulations are strict

***examiner's* note** Sometimes the term 'ecotourism' is applied wrongly to attract business, where profit is the main motive.

National Parks

Q1 Give the two main reasons why National Parks were established in the UK.

Q2 State three problems or conflicts in the UK's National Parks.

Q3 Name one of the three new National Parks designated in the UK since 2000.

Q4 In what ways are the locations of the three new National Parks different from those of the old ones?

A1 to preserve beautiful natural landscapes; to provide facilities for visitors

A2 visitor pressure in the most popular places ('honeypots'); footpath erosion; conflicts with farmers over open gates, litter etc.; conflicts with village residents over noise and buying up houses at high prices; traffic congestion on weekends

A3 the Cairngorms, Loch Lomond and the New Forest; soon the South Downs

A4 the new parks include Scotland's first two National Parks, and the New Forest which is in the south of England; most of the old parks are in the uplands in the north and west of England and Wales

***examiner's* note** National Parks have been set up in many other countries for similar reasons.

Fossil fuels

Q1 Name the three fossil fuels.

Q2 Explain why fossil fuels are examples of non-renewable resources.

Q3 About 75% of world energy supplies come from fossil fuels. Gives reasons why this percentage is so high.

Q4 Name two types of atmospheric pollution caused by fossil fuels.

ANSWERS

energy sources from plants and animals that died millions of years ago

A1 coal, oil and natural gas

A2 they are natural resources that can be used only once; it will take millions of years for new resources to form; old resources are being consumed much more quickly than new resources can be formed

A3 they are easy to use, have high heating power, are cheaper than other energy sources and are versatile fuels that can be used in different ways; technology was developed based on these energy sources

A4 smog in urban areas; acid rain; enhanced greenhouse effect leading to global warming

examiner's note A world without fossil fuels is difficult to imagine, but known oil reserves could last for only around 40–60 years, and will not be as cheap as in the past.

Alternative energy sources

Q1 Name three alternative sources of energy.

Q2 Explain why these alternative energy sources are examples of renewable resources.

Q3 Alternative energy sources account for less than 10% of world energy production. Why is this percentage so low?

Q4 Why does the UK government want to increase the percentage of energy from alternative sources?

renewable means of generating electricity instead of burning fossil fuels

A1 hydroelectric power (HEP), wind, solar, geothermal and tidal

A2 they depend on naturally-occurring features which are constantly present, such as weather and water, so they will never run out

A3 the world is set up to use fossil fuels; only HEP is a cheap, well-established alternative source of electricity; research and development costs for the others are high; not only does electricity from them cost more but it cannot be guaranteed, e.g. days without wind or sun

A4 to reduce emissions of greenhouse gases like carbon dioxide; to lower air pollution in cities

***examiner's* note** Progress towards the UK's 15% target from renewable energy sources by 2020 has been slow and relies mostly on wind power.

World development

Q1 Describe the course taken by the development line on the map.

Q2 What are the main differences in economic development between countries lying north and south of this line?

A1 the line mainly runs east to west, except off eastern Asia where it bends south to include Oceania on its northern side; countries in North America and Europe lie to the north whereas South American and African countries lie to the south; countries in Asia lie on both sides

A2 MEDCs are to the north of the line, with high incomes per head and are industrialised; LEDCs are to the south of the line, with low incomes and more reliance on farming

***examiner's* note** Although very simple, this map makes a useful two-fold division of the world based on development and wealth.

Measures of development

Q1 What is GDP and how is it calculated?

Q2 State one advantage and one disadvantage of GDP as a measure of development.

Q3 What does the term 'infant mortality rate' mean? Why is it considered to be a reliable measure of development?

Q4 Name examples of other social measures of development related to health and education.

these include both economic and social indicators

A1 gross domestic product: total value of all goods and services produced in a country in a year

A2
- advantage: can make it possible to compare wealth of countries
- disadvantage: impossible to be accurate in countries with many subsistence workers

A3
- the number of deaths of children under 1 year old per 1000 people
- a high rate shows that medical care and food supplies are inadequate, because babies are most vulnerable

A4 life expectancy; number of people per doctor; percentage of people without access to clean water; average calorie intake per day; adult literacy rate; proportion of children who attend school

***examiner's* note** A combined term 'socioeconomic' is often used.

Aid

Q1 Describe the characteristics of short-term aid.

Q2 In what ways is long-term aid different?

Q3 Some aid is described as 'tied aid'. What does this mean?

Q4 What are NGOs? Some people regard aid from NGOs as the best type of aid. Explain why.

money or help given to a country in need

A1 emergency aid of food, medicines and shelter, often after a natural disaster like an earthquake

A2 long-term aid is development aid to improve quality of life, such as help with digging wells and training farmers

A3 aid-giving governments decide how it will be spent; often it is equipment for large projects like dams and power stations

A4
- non-governmental organisations, mainly charities such as Save the Children
- NGOs fund small-scale projects that benefit local people, like building clinics, schools and wells

***examiner's* note** Aid must not be seen as a solution to the problem of low economic development. As a temporary measure it saves lives and relieves suffering.

Globalisation

Q1 There has been a 'revolution in global communications'. Explain why it is now easier than ever before for businesses and people in different parts of the world to keep in touch with each other, and for people to travel between countries.

Q2 What is a container ship? Describe how these ships allow easier export and import of manufactured goods.

Q3 Explain how transnational corporations contribute to globalisation.

the increasing importance of international operations for businesses and people

A1 • ease of keeping in touch: phone, fax, internet and e-mail, texting; for businesses, video conferencing
- ease of travel: most places in the world can be reached in 24 hours by aeroplane

A2 a ship carrying metal boxes of standard sizes; the box is packed with goods (e.g. trainers) and sealed in the factory, carried by lorry and ship and not re-opened until the country of sale is reached

A3 factories are set up where costs of production are lowest; world location does not matter; transport of goods by container ship is relatively cheap

***examiner's* note** Globalisation means that people are affected by business decisions made thousands of miles away, over which they have no control.

Interdependence

Q1 State two advantages resulting from trade between countries.

Q2 Describe the main patterns of world trade between developed and developing countries.

Q3 Why do these patterns of world trade favour developed countries?

Q4 What is being done to make trade and interdependence between countries fairer to less developed countries (LEDCs)?

when countries exchange goods and services that they cannot supply for themselves

A1 surplus products sold; country can specialise in what it does best; it can obtain goods it cannot make or grow itself

A2 developed countries export mainly technology, machinery and high-value manufactured goods; developing countries export mainly primary products (minerals and crops)

A3 prices of primary products from developing countries are low and fluctuate; prices of machinery are high as companies in developed countries expect a good profit

A4 the World Trade Organization promotes free trade for products from LEDCs; Fair trade organisations pay higher than world prices to some producers, e.g. coffee growers, and support local communities

***examiner's* note** Any improvements in trade for developing countries are hard won.

Sustainable development

Q1 How does the item shown in the photo contribute to sustainable development?

Q2 Why do some people disapprove of wind turbines?

Q3 Other suitable energy sources are solar, water (HEP), wave and tidal power. What do these have in common?

ANSWERS

activities for economic growth that involve working with the environment for a long-term future

A1 wind turbines generate electricity using the weather, a never-ending natural resource; no atmospheric pollution results; fossil fuels are not used up; it is a renewable resource so the Earth will remain the same for future generations

A2 some people complain about noise, spoilt views in scenic countryside, interference with flight paths of birds and with television reception; there is no power on calm days, meaning that fossil fuel power stations are still needed; wind turbines are a more expensive form of energy than fossil fuels

A3 they are renewable and obtained from natural resources

***examiner's* note** Governments and businesses are under increasing pressure to implement options for development that are sustainable.